Table of Content

Chapter 1: Introduction

1.1 Importance of Transistor Testing and Characterization

Transistors are fundamental building blocks of modern electronic systems, and their performance and reliability are crucial for the overall functionality of these systems. Testing and characterization of transistors are essential steps in the design and manufacturing processes to ensure that the transistors meet the required specifications and perform as expected.

Quality Control and Process Monitoring

Transistor testing plays a vital role in quality control during manufacturing. By performing electrical tests on each transistor, manufacturers can identify and eliminate defective devices. This helps to ensure that only high-quality transistors are used in electronic products, reducing the risk of failures and improving the reliability of the end products.

Furthermore, transistor testing provides valuable data for process monitoring and improvement. By analyzing the test results, manufacturers can identify trends and variations in transistor performance. This information can help to identify areas for process optimization, leading to higher yields and better overall quality.

Device Characterization and Modeling

Transistor characterization involves measuring and analyzing the electrical properties of transistors under various operating conditions. This data is used to develop accurate models that describe the transistor's behavior. These models are essential for circuit design and simulation, allowing engineers to predict and optimize the performance of electronic circuits.

Accurate characterization is also crucial for advanced transistor technologies, such as high-frequency transistors and power transistors. By understanding the limitations and capabilities of these devices, designers can push the boundaries of electronic system performance.

Failure Analysis and Root Cause Identification

Transistor testing is also a valuable tool for failure analysis and root cause identification. When an electronic system fails, testing individual transistors can help to identify the faulty device and determine the cause of failure. This information is critical for troubleshooting and preventing future failures.

By performing root cause analysis, engineers can identify

design or manufacturing defects that may have contributed to the failure. This knowledge can lead to improvements in device design, manufacturing processes, or system-level design to prevent similar failures in the future.

Cost-Effective Manufacturing

Transistor testing and characterization contribute to cost-effective manufacturing. By identifying defective devices early in the production process, manufacturers can avoid costly rework or scrappage. Additionally, by optimizing transistor performance through characterization and modeling, manufacturers can reduce the need for overdesign and improve the efficiency of their products.

Conclusion

In summary, transistor testing and characterization are essential processes for ensuring the quality, performance, and reliability of electronic systems. By performing electrical tests, analyzing the data, and developing accurate models, engineers and manufacturers can identify and mitigate potential issues, optimize device performance, and troubleshoot failures. The importance of transistor testing and characterization cannot be overstated in the design, manufacturing, and operation of modern electronic systems.

1.2 Overview of Transistor Technologies

1. 2. 1 Introduction to Transistors

Transistors, the fundamental building blocks of modern electronics, are semiconductor devices that regulate the flow of electrical current. They operate as electronic switches, allowing for precise control over the flow of electrons within electronic circuits. Since their invention in the mid-20th century, transistors have revolutionized the field of electronics, enabling the miniaturization and exponential growth of computing power.

1. 2. 2 Basic Transistor Structure and Operation

Transistors are typically constructed from semiconductor materials, such as silicon or gallium arsenide. The basic structure of a transistor consists of three terminals: the emitter, base, and collector. When a voltage is applied between the base and emitter terminals, it creates an electric field that either attracts or repels electrons within the device. This, in turn, controls the flow of current between the emitter and collector terminals.

1. 2. 3 Types of Transistors

There are two main types of transistors: bipolar junction transistors (BJTs) and metal-oxide-semiconductor field-effect transistors (MOSFETs). BJTs utilize the movement of both electrons and holes (the absence of electrons) to control current flow, while MOSFETs rely solely on electrons. Each type has its unique advantages and applications.

1. 2. 4 Bipolar Junction Transistors (BJTs)

BJTs are characterized by their three semiconductor layers: an emitter, base, and collector. Current flow in a BJT is controlled by the base-emitter junction, which acts as a voltage-controlled switch. When a positive voltage is applied to the base, it attracts electrons from the emitter, allowing current to flow between the emitter and collector.

1. 2. 5 Metal-Oxide-Semiconductor Field-Effect Transistors (MOSFETs)

MOSFETs are distinguished by their insulated gate, which separates the controlling electrode from the semiconductor channel. Unlike BJTs, MOSFETs use an electric field to control the flow of electrons between the source and drain terminals. When a positive voltage is applied to the gate, it attracts electrons, forming a conductive channel that allows current to flow.

1. 2. 6 Transistor Applications

Transistors find applications in a vast array of electronic devices, including computers, smartphones, televisions, and industrial control systems. They serve as the essential components in digital logic circuits, amplifiers, oscillators, and many other electronic functions.

1. 2. 7 Scaling and Moore's Law

The miniaturization of transistors has been a key driver of technological advancements over the past several decades. Moore's Law, proposed by Intel co-founder Gordon Moore in 1965, states that the number of transistors on a computer chip doubles approximately every two years. This exponential growth has led to the continuous improvement in computational power and the development of increasingly sophisticated electronic devices.

1. 2. 8 Emerging Transistor Technologies

Ongoing research and development efforts focus on exploring new transistor technologies to overcome the limitations of conventional silicon-based devices. These emerging technologies include compound semiconductors, graphene, and nanowires, which promise enhanced performance, reduced power consumption, and further miniaturization.

1. 2. 9 Conclusion

Transistors are indispensable components of modern electronics, providing the essential functionality for controlling and amplifying electrical signals. The ongoing evolution of transistor technologies continues to drive technological advancements and shape the future of electronics.

1.3 Challenges in Modern Transistor Testing

Testing transistors accurately and efficiently is crucial to ensure the reliability and performance of these devices. However, the rapid advancement of transistor technology has posed significant challenges to traditional testing methods. This section examines the key challenges in modern transistor testing and explores innovative approaches to overcome these limitations.

Challenges in Scaling and Miniaturization
As the demand for smaller, faster, and more energy-efficient devices increases, transistors are being scaled down to nanoscale dimensions. This miniaturization has led to several challenges:

Increased Parasitics: The smaller size of transistors results in increased parasitic capacitance and resistance, which can affect the accuracy of test measurements and limit the device's performance.
Reduced Signal-to-Noise Ratio (SNR): The reduced size of transistors also decreases the signal levels, making it more challenging to distinguish between the actual signal and noise during testing.

Leakage Currents and Reliability
Modern transistors exhibit significant leakage currents due to various factors such as quantum tunneling and gate oxide thinning. These leakage currents can lead to increased power consumption and reduced device

reliability. Testing leakage currents accurately is crucial to ensure the longevity and energy efficiency of electronic devices.

Variability and Variability Testing

Transistors manufactured using advanced fabrication processes exhibit significant variability in their electrical characteristics. This variability stems from variations in material properties, process conditions, and device geometry. Accurate testing is necessary to identify and characterize this variability and ensure that devices meet performance specifications.

High-Speed and RF Testing

The increasing use of high-speed and radio frequency (RF) devices in modern electronics has necessitated the development of specialized testing techniques. These devices operate at frequencies exceeding gigahertz (GHz), requiring test equipment with high bandwidth and low noise. Traditional testing methods may not be adequate for characterizing the performance of these high-speed and RF devices.

Thermal Considerations

Transistors generate heat during operation, which can affect their electrical characteristics. Accurate testing requires controlling the temperature of the device during testing to prevent self-heating effects from influencing the measurement results.

Innovative Approaches to Modern Transistor Testing

Advanced Measurement Techniques:
Pulsed I-V Testing: Uses short pulses to minimize self-heating effects and improve SNR.
Capacitance-Voltage (C-V) Profiling: Characterizes the gate oxide properties and detects defects.
Noise Spectroscopy: Analyzes the noise power spectrum to identify and diagnose specific defects.

Parametric Test Optimization:
Artificial Intelligence (AI): Uses machine learning algorithms to optimize test parameters and improve test accuracy.
Adaptive Testing: Adjusts test parameters based on real-time device feedback to improve test efficiency and coverage.

Variability Characterization and Modeling:
Statistical Sampling: Uses statistical techniques to characterize variability and identify outliers.
Variability Modeling: Develops models to predict and compensate for device variability.

High-Speed and RF Testing:
Vector Network Analyzers (VNA): Measures the scattering parameters of RF devices to characterize their impedance and frequency response.
Signal Generators and Analyzers: Generate and analyze high-frequency signals for testing high-speed devices.

Thermal Management:

Temperature-Controlled Test Fixtures: Maintain the device temperature at a specific setpoint during testing. Thermal Imaging: Monitors the temperature distribution of the device under test to identify potential hotspots.

Conclusion

Modern transistor testing faces significant challenges due to scaling, miniaturization, leakage currents, variability, high-speed operation, and thermal considerations. Overcoming these challenges requires innovative approaches such as advanced measurement techniques, parametric test optimization, variability characterization, high-speed and RF testing, and thermal management. These advancements enable accurate and efficient testing of modern transistors, ensuring the reliability, performance, and longevity of electronic devices.

Chapter 2: Fundamentals of Transistors

2.1 Basic Transistor Operation and Characteristics

Transistors are semiconductor devices that act as electronic switches or amplifiers. They are the fundamental building blocks of modern electronics and play a crucial role in shaping the technological landscape of today. To comprehend the intricacies of electronics, it is imperative to grasp the basic principles governing the operation and characteristics of transistors.

2. 1. 1 Structure and Function of a Transistor

A transistor, typically fabricated from silicon, consists of three terminals: emitter, base, and collector. These terminals are connected to semiconductor regions called emitter, base, and collector, respectively. The emitter and collector regions are separated by a thin base region, forming two junctions: the emitter-base junction and the collector-base junction.

In its simplest form, a transistor can be visualized as a switch that can be either turned on or off. When a small voltage is applied between the base and emitter, the emitter-base junction becomes forward biased, allowing electrons to flow from the emitter to the base. This flow of electrons modulates the resistance between the collector and emitter, enabling the transistor to act as a switch or an amplifier.

2. 1. 2 Transistor Characteristics

The behavior of a transistor is characterized by a set of parameters that define its electrical properties. These parameters include:

- Current Gain (β): The ratio of collector current to base current, indicating the transistor's ability to amplify current.

- Transconductance (gm): The ratio of collector current to base-emitter voltage, representing the transistor's ability to convert voltage changes into current changes.

- Cutoff Frequency (fT): The frequency at which the transistor's current gain falls to unity, limiting the transistor's high-frequency performance.

- Input Resistance (rin): The resistance between the base and emitter terminals, influencing the transistor's sensitivity to input signals.

- Output Resistance (rout): The resistance between the collector and emitter terminals, affecting the transistor's ability to drive external loads.

2. 1. 3 Transistor Configurations

Transistors can be connected in three basic configurations: common-emitter, common-base, and common-collector. Each configuration offers distinct advantages and is suitable for specific applications.

- Common-Emitter (CE): Provides high current gain and voltage amplification, making it ideal for signal amplification and power switching.

- Common-Base (CB): Offers high input impedance and low output impedance, suitable for impedance matching and voltage buffering.

- Common-Collector (CC): Provides high input impedance and unity voltage gain, acting as a current amplifier or voltage follower.

2. 1. 4 Transistor Applications

Transistors find widespread application in a vast array of electronic circuits, including:

- Amplifiers: Amplify weak signals for applications such as audio systems and telecommunications.

- Switches: Control the flow of current in digital circuits and power electronics.

- Oscillators: Generate periodic waveforms for use in

timing circuits and communication systems.

- Logic Gates: Implement basic logic operations in digital circuits, forming the foundation of computer systems.

- Integrated Circuits (ICs): Combine multiple transistors and other electronic components on a single chip, enabling the realization of complex electronic functions.

2. 1. 5 Summary

Transistors are versatile electronic devices that serve as the cornerstone of modern electronics. Understanding their basic operation and characteristics is essential for comprehending the behavior and design of electronic circuits. The three fundamental transistor configurations, combined with their unique electrical properties, enable transistors to perform a wide range of functions, from signal amplification to logic operations.

2.2 Transistor Models and Parameters

Transistors are the fundamental building blocks of modern electronics, and their behavior is crucial to understanding the operation of electronic circuits. To accurately predict and analyze circuit behavior, it is essential to have accurate models of transistors. In this section, we will explore the various transistor models and parameters that are commonly used in circuit analysis and design.

2. 2. 1 Transistor Models

Transistor models are mathematical representations of the electrical behavior of transistors. They simplify the complex physical processes that occur within the transistor into a set of equations and parameters that can be used in circuit analysis. The most common transistor models include:

Ideal Transistor Model: The ideal transistor model assumes that the transistor is either fully ON (conducting) or fully OFF (non-conducting). This model is simple and useful for basic circuit analysis, but it does not accurately represent the actual behavior of transistors.
Linear Model: The linear model assumes that the transistor operates in a linear region where the relationship between the input and output voltages and currents is linear. This model is useful for analyzing small-signal circuits, where the transistor operates within a narrow range of its operating point.
Nonlinear Model: The nonlinear model takes into account the nonlinear behavior of transistors, which occurs at higher operating currents and voltages. This model is more accurate than the linear model, but it is also more complex to use.

2. 2. 2 Transistor Parameters

Transistor parameters are the numerical values that characterize the electrical behavior of transistors.

These parameters are used in transistor models to predict the transistor's performance in a given circuit. The most common transistor parameters include:

β (Beta): The forward current gain of the transistor, which is the ratio of the collector current to the base current.
α (Alpha): The common-base current gain of the transistor, which is the ratio of the collector current to the emitter current.
hfe: The forward current gain of the transistor in the common-emitter configuration.
hfe: The reverse current gain of the transistor in the common-collector configuration.
Vbe: The base-emitter voltage, which is the voltage drop between the base and emitter terminals.
Vce: The collector-emitter voltage, which is the voltage drop between the collector and emitter terminals.
Ic: The collector current, which is the current flowing through the collector terminal.
Ib: The base current, which is the current flowing through the base terminal.
Ie: The emitter current, which is the current flowing through the emitter terminal.

2. 2. 3 Transistor Biasing

Transistor biasing is the process of setting the operating point of a transistor, which is the point at which the transistor operates in a stable and predictable manner. Biasing is necessary to ensure that the transistor operates in the desired region (linear or nonlinear) and that it is able to amplify signals effectively.

The most common biasing methods include:

Fixed Bias: In fixed bias, a constant voltage is applied to the base terminal of the transistor, which sets the operating point.
Emitter Bias: In emitter bias, a resistor is connected between the emitter terminal and the ground, which stabilizes the operating point.
Collector Feedback Bias: In collector feedback bias, a resistor is connected between the collector terminal and the base terminal, which provides negative feedback and stabilizes the operating point.

2. 2. 4 Transistor Applications

Transistors are used in a wide range of electronic applications, including:

Amplifiers: Transistors can be used to amplify signals, increasing their voltage, current, or power.
Switches: Transistors can be used as switches, turning on or off the flow of current in a circuit.
Logic Gates: Transistors can be used to implement logic gates, which are the basic building blocks of digital circuits.
Oscillators: Transistors can be used to generate oscillating signals, which are used in a variety of applications, such as clocks and signal generators.

2.3 Measurement Techniques and Standards

In order to ensure accurate and reliable measurements, it is crucial to adhere to standardized measurement techniques and standards. These guidelines provide a framework for conducting measurements consistently and minimizing errors.

Measurement Techniques

There are various measurement techniques employed in scientific and engineering disciplines, each with its own advantages and applications. Some of the most common techniques include:

Direct Measurement: This involves directly measuring a quantity using a measuring instrument, such as a ruler, thermometer, or balance. Direct measurement is typically straightforward but can be limited in precision and accuracy.

Indirect Measurement: This method involves using a relationship between the measured quantity and another variable that can be measured more easily. For instance, measuring the resistance of a resistor using a voltmeter and Ohm's law. Indirect measurement allows for greater precision but requires careful calibration and analysis.

Comparative Measurement: This technique compares the measured quantity to a known standard. For example, using a balance to compare the weight of an object to a calibrated set of weights. Comparative measurement is often used for high-precision measurements.

Measurement Standards

Measurement standards serve as references for calibrating measuring instruments and ensuring the accuracy and consistency of measurements across different devices and laboratories. These standards are typically established by national or international organizations and are based on fundamental physical constants or well-defined physical phenomena.

Some of the most important measurement standards include:

International System of Units (SI): SI is the modern form of the metric system and defines the seven base units (meter, kilogram, second, ampere, kelvin, mole, and candela) and their derived units.

National Institute of Standards and Technology (NIST): NIST is the United States' national measurement standards laboratory and provides traceable calibration services and reference materials for various measurement disciplines.

International Organization for Standardization (ISO): ISO develops and publishes international standards for a wide range of industries, including those related to measurement and calibration.

Importance of Measurement Techniques and Standards

Adhering to standardized measurement techniques and standards is essential for several reasons:

Accuracy and Reliability: Standardized techniques and standards minimize measurement errors and ensure that measurements are accurate and reliable.

Consistency and Traceability: Using standardized methods and calibrating instruments against certified standards ensures that measurements are consistent and can be traced back to a recognized reference.

Comparability: Standardization enables the comparison of measurements made by different researchers, laboratories, and industries, facilitating collaboration and knowledge sharing.

Quality Control and Assurance: Standardized measurement practices are crucial for quality control and assurance in various fields, such as manufacturing, healthcare, and environmental monitoring.

Safety and Regulatory Compliance: Compliance with measurement standards is often required for safety and regulatory purposes, ensuring that products and processes meet specified requirements.

Conclusion

Measurement techniques and standards are fundamental to ensuring the accuracy, reliability, consistency, and comparability of measurements. By adhering to standardized procedures and calibrating against recognized references, scientists, engineers, and professionals can obtain meaningful and trustworthy measurement results that contribute to advancements in various fields and disciplines.

Chapter 3: Electrical Testing Techniques

3.1 DC Characterization: IV Curves and Parameter Extraction

The electrical characteristics of a semiconductor device can be effectively captured through DC characterization, a fundamental step in device modeling and performance assessment. DC characterization involves measuring the device's current-voltage (IV) curves and extracting key electrical parameters that describe its behavior under DC operating conditions. These parameters provide insights into the device's functionality, limitations, and potential applications.

IV Curves

IV curves are graphical representations of the relationship between the voltage applied to a device and the resulting current that flows through it. By varying the applied voltage and measuring the corresponding current, a family of IV curves can be obtained. These curves provide a comprehensive view of the device's electrical behavior over a range of operating conditions.

The shape and characteristics of the IV curves depend on the type of semiconductor device being characterized.

For example, a typical diode IV curve exhibits a forward-bias region, where current increases exponentially with increasing voltage, and a reverse-bias region, where current remains relatively low. In contrast, a typical transistor IV curve displays three distinct regions: cutoff, where no current flows; saturation, where current remains relatively constant; and active, where current is linearly proportional to voltage.

Parameter Extraction

From the IV curves, various electrical parameters can be extracted to quantify the device's performance. These parameters include:

Threshold voltage (Vth): The minimum gate voltage required to turn on a transistor.
Transconductance (gm): The ratio of output current to input voltage in the active region of a transistor.
Output resistance (ro): The ratio of output voltage to output current in the saturation region of a transistor.
Diode ideality factor (n): A measure of the deviation of a diode's IV curve from the ideal exponential behavior.

Parameter extraction techniques involve curve fitting and mathematical analysis to determine these parameters accurately. The extracted parameters provide valuable information for circuit design, device modeling, and performance optimization.

Applications

DC characterization plays a crucial role in various aspects of semiconductor device development and utilization:

Device Modeling: The extracted parameters form the basis for developing accurate device models that can be used in circuit simulations and design tools.
Process Monitoring: DC characterization is used to monitor the manufacturing process and ensure consistency in device performance.
Failure Analysis: IV curves can be used to diagnose device failures and identify potential defects.
Circuit Design: The extracted parameters guide the selection of appropriate devices for specific circuit applications.

Conclusion

DC characterization is an essential technique for understanding and quantifying the electrical behavior of semiconductor devices. Through the analysis of IV curves and parameter extraction, valuable insights can be gained into the device's functionality, limitations, and potential applications. DC characterization plays a critical role in device modeling, process monitoring, failure analysis, and circuit design, enabling the development and optimization of high-performance electronic systems.

3.2 AC Characterization: Frequency Response and Small-Signal Parameters

The frequency response of a transistor describes how its electrical characteristics vary with frequency. It is an important consideration for high-frequency applications, such as RF amplifiers and oscillators. The small-signal parameters of a transistor are a set of parameters that describe its behavior under small-signal conditions. They are used to design and analyze transistor circuits.

Frequency Response

The frequency response of a transistor is typically characterized by its cutoff frequencies. The cutoff frequencies are the frequencies at which the transistor's gain drops by 3 dB. The low-frequency cutoff frequency is typically determined by the transistor's input capacitance, while the high-frequency cutoff frequency is typically determined by the transistor's output capacitance.

The frequency response of a transistor can be measured using a frequency response analyzer. A frequency response analyzer is a test instrument that measures the amplitude and phase of a signal over a range of frequencies.

Small-Signal Parameters

The small-signal parameters of a transistor are a set of parameters that describe its behavior under small-signal conditions. They are used to design and analyze transistor circuits. The small-signal parameters of a transistor are typically measured using a small-signal parameter analyzer. A small-signal parameter analyzer is a test instrument that measures the small-signal parameters of a transistor.

The small-signal parameters of a transistor include:

Input resistance (rπ): The input resistance is the resistance of the transistor's input terminal when it is biased in the active region.
Output resistance (r0): The output resistance is the resistance of the transistor's output terminal when it is biased in the active region.
Forward current gain (β): The forward current gain is the ratio of the collector current to the base current when the transistor is biased in the active region.
Reverse voltage gain (α): The reverse voltage gain is the ratio of the collector voltage to the base voltage when the transistor is biased in the active region.

The small-signal parameters of a transistor can be used to design and analyze transistor circuits. For example, the input resistance can be used to determine the input impedance of a transistor amplifier, and the output resistance can be used to determine the output impedance of a transistor amplifier.

Example

The following example shows how to use the small-signal parameters of a transistor to design a transistor amplifier.

Problem:

Design a transistor amplifier with a voltage gain of 10.

Solution:

The voltage gain of a transistor amplifier is given by the following equation:

```
Av = -β RL . (rπ + RB)
```

where:

Av is the voltage gain
β is the forward current gain of the transistor
RL is the load resistance
rπ is the input resistance of the transistor
RB is the bias resistor

We want the voltage gain to be 10, so we can solve the above equation for RL:

```
RL = 10 (rπ + RB) . β
```

We know that the input resistance of the transistor is 1

kΩ and the bias resistor is 10 kΩ. We also know that the forward current gain of the transistor is 100.
Substituting these values into the above equation, we get:

```
RL = 10 (1 kΩ + 10 kΩ) . 100 = 110 kΩ
```

Therefore, we need to use a load resistor of 110 kΩ to achieve a voltage gain of 10.

Conclusion

The frequency response and small-signal parameters of a transistor are important considerations for high-frequency applications and for designing transistor circuits. The frequency response of a transistor can be measured using a frequency response analyzer, and the small-signal parameters of a transistor can be measured using a small-signal parameter analyzer.

3.3 Noise Measurements and Analysis

Noise is an unwanted sound that can interfere with communication, sleep, and other activities. It can also be harmful to our health, causing hearing loss, cardiovascular disease, and other problems.

There are many different ways to measure noise. The most common method is to use a sound level meter. This device measures the sound pressure level (SPL) in decibels (dB). SPL is a logarithmic scale that measures the relative loudness of a sound. A sound level of 0 dB is the threshold of human hearing, while a sound level of 120 dB can cause pain.

In addition to SPL, there are other factors that can affect the perceived loudness of a sound. These factors include the frequency of the sound, the duration of the sound, and the background noise level.

Frequency is measured in hertz (Hz). The human ear is most sensitive to sounds in the mid-frequency range (500-2000 Hz). Sounds at higher or lower frequencies are not as loud.

Duration is measured in seconds. A short-duration sound is less annoying than a long-duration sound.

Background noise is the noise that is always present in the environment. The background noise level can affect the perceived loudness of a sound. A sound that is louder than the background noise will be more noticeable than a sound that is quieter than the background noise.

Noise measurements can be used to assess the noise impact of a particular activity or to determine compliance with noise regulations. Noise measurements can also be used to identify noise sources and to develop noise control strategies.

There are a number of different ways to analyze noise

measurements. One common method is to use a noise histogram. A noise histogram shows the distribution of sound levels over time. This information can be used to identify the peak sound levels and the average sound levels.

Another common method of analyzing noise measurements is to use a frequency analysis. A frequency analysis shows the sound pressure level at different frequencies. This information can be used to identify the dominant frequencies in a noise source.

Noise measurements and analysis can be a valuable tool for assessing the noise impact of a particular activity or for determining compliance with noise regulations. Noise measurements can also be used to identify noise sources and to develop noise control strategies.

3. 3. 1 Noise Measurement Instruments

There are a variety of noise measurement instruments available, each with its own advantages and disadvantages. The most common type of noise measurement instrument is the sound level meter. Sound level meters are relatively inexpensive and easy to use, and they can provide accurate measurements of sound pressure level.

Other types of noise measurement instruments include noise dosimeters, noise analyzers, and sound level calibrators. Noise dosimeters are used to measure the cumulative noise exposure of a worker over a period of time. Noise analyzers are used to measure the frequency content of a noise source. Sound level calibrators are used to calibrate sound level meters.

3. 3. 2 Noise Measurement Techniques

There are a number of different noise measurement techniques that can be used to measure the noise impact of a particular activity. The most common noise measurement technique is the short-term measurement. Short-term measurements are typically taken over a period of 10-15 minutes.

Other noise measurement techniques include the long-term measurement and the statistical measurement. Long-term measurements are typically taken over a period of 24 hours or more. Statistical measurements are used to determine the distribution of sound levels over time.

3. 3. 3 Noise Analysis Techniques

There are a number of different noise analysis techniques that can be used to analyze noise measurements. The most common noise analysis technique is the noise histogram. A noise histogram shows the distribution of sound levels over time. This information can be used to identify the peak sound levels and the average sound levels.

Other noise analysis techniques include the frequency analysis and the time-history analysis. A frequency analysis shows the sound pressure level at different frequencies. This information can be used to identify the dominant frequencies in a noise source. A time-history analysis shows the sound pressure level over time. This information can be used to identify the time-varying characteristics of a noise source.

3. 3. 4 Noise Control Strategies

There are a number of different noise control strategies that can be used to reduce the noise impact of a particular activity. The most common noise control strategy is to use sound barriers. Sound barriers can be used to block the path of sound waves and reduce the sound pressure level at a particular location.

Other noise control strategies include the use of sound absorbers, sound dampers, and vibration isolators. Sound absorbers are used to absorb sound waves and reduce the reverberation time in a room. Sound dampers are used to reduce the transmission of sound waves through a wall or other barrier. Vibration isolators are used to reduce the transmission of vibration from a machine or other source to a structure.

The selection of the appropriate noise control strategy depends on the specific noise problem and the desired noise reduction goal.

Chapter 4: Reliability Testing and Characterization

4.1 Stress-Induced Degradation and Failure Mechanisms

Stress-induced degradation (SID) refers to the gradual deterioration of a material's properties and performance under the influence of external stressors. These stressors can be mechanical (e. g. , loading, vibration), thermal (e. g. , temperature fluctuations), chemical (e. g. , corrosion), or electrical (e. g. , voltage surges). Over time, SID can lead to the failure of the material or component.

SID occurs through various mechanisms, including:

Mechanical fatigue: Repeated loading and unloading cycles cause the formation and propagation of cracks, ultimately leading to failure.
Creep: Prolonged exposure to high temperatures and stresses causes gradual deformation and weakening of the material.
Stress corrosion cracking (SCC): The combined action of stress and a corrosive environment leads to the formation and growth of cracks, compromising the material's integrity.
Thermal fatigue: Repeated heating and cooling cycles

cause thermal stresses that can lead to crack initiation and propagation.
Electromigration: The movement of ions within a conductor under the influence of an electric field can cause the formation of voids and ultimately lead to failure.

Understanding SID mechanisms is crucial for predicting the reliability and lifespan of materials and components. It enables engineers to design and implement strategies to mitigate SID and prevent premature failure.

Fatigue Failure

Fatigue failure is a common mode of failure in materials subjected to cyclic loading. It occurs when the material experiences a series of repetitive stresses below its ultimate tensile strength, eventually leading to crack initiation and propagation. Fatigue failure is particularly prevalent in components that undergo high-cycle loading, such as aircraft wings, automotive suspension systems, and electronic devices.

The fatigue life of a material is influenced by several factors, including:

Stress amplitude: The magnitude of the applied stress.
Stress ratio: The ratio of minimum to maximum stress during the loading cycle.
Frequency: The number of loading cycles per unit time.
Material properties: The inherent strength and

toughness of the material.

Fatigue failure can be prevented or delayed by employing various techniques, such as:

Stress reduction: Reducing the applied stress levels.
Shot peening: Introducing compressive stresses on the surface of the material to inhibit crack formation.
Fatigue-resistant materials: Using materials with high fatigue strength and crack resistance.

Creep Failure

Creep failure occurs when a material subjected to constant stress at elevated temperatures undergoes gradual deformation and weakening. This is a common issue in components operating at high temperatures, such as turbine blades, pressure vessels, and pipelines.

Creep failure mechanisms involve the movement of dislocations, grain boundary sliding, and diffusion processes. These mechanisms lead to the accumulation of strain over time, eventually resulting in failure.

Factors influencing creep failure include:

Stress level: The higher the stress, the faster the creep rate.
Temperature: Elevated temperatures accelerate creep processes.
Time: Creep failure occurs over extended periods of

exposure to stress and temperature.
Material properties: Materials with high creep resistance are less susceptible to failure.

Creep failure can be mitigated by:

Reducing stress levels: Lowering the operating stress.
Using creep-resistant materials: Selecting materials with high creep strength and stability.
Annealing: Periodic heat treatments to relieve accumulated stresses.

Stress Corrosion Cracking

Stress corrosion cracking (SCC) is a failure mechanism that occurs when a material is simultaneously subjected to tensile stress and a corrosive environment. The presence of corrosive agents, such as chlorides, sulfides, or hydrogen, enhances crack initiation and propagation.

SCC is a major concern in industries that involve exposure to corrosive environments, such as the oil and gas, chemical processing, and nuclear power industries.

Factors influencing SCC include:

Stress level: The higher the stress, the greater the risk of SCC.
Corrosive environment: The presence and concentration of corrosive agents.
Material properties: Materials with high SCC resistance

are less susceptible to failure.

Mitigating SCC involves:

Stress reduction: Reducing the applied stress levels.
Corrosion protection: Using protective coatings or barriers to shield the material from the corrosive environment.
Material selection: Selecting materials with high SCC resistance.

Thermal Fatigue Failure

Thermal fatigue failure occurs when a material is subjected to repeated heating and cooling cycles, causing thermal stresses that lead to crack formation and propagation. This is a common failure mode in components exposed to extreme temperature fluctuations, such as combustion engines, electronic components, and thermal processing equipment.

Factors influencing thermal fatigue failure include:

Temperature range: The magnitude of the temperature variations.
Cycling frequency: The number of heating and cooling cycles.
Material properties: Materials with high thermal conductivity and low coefficient of thermal expansion are less susceptible to thermal fatigue.

Mitigating thermal fatigue failure involves:

Reducing temperature variations: Minimizing the temperature range experienced by the component.
Slowing down thermal cycling: Reducing the cycling frequency.
Using thermal fatigue-resistant materials: Selecting materials with high thermal conductivity and low coefficient of thermal expansion.

Electromigration Failure

Electromigration is a failure mechanism that occurs in conductors when electric current causes the movement of ions. This movement can lead to the formation of voids and hillocks, which can disrupt the electrical conductivity of the conductor.

Electromigration is a major concern in high-density integrated circuits (ICs) and other electronic devices where current densities are high.

Factors influencing electromigration failure include:

Current density: The higher the current density, the greater the risk of electromigration.
Temperature: Elevated temperatures accelerate electromigration processes.
Material properties: Materials with low electromigration resistance are more susceptible to failure.

Mitigating electromigration failure involves:

Reducing current densities: Decreasing the current flowing through the conductors.
Using electromigration-resistant materials: Selecting materials with high electromigration resistance.
Implementing design techniques: Employing design rules and layout techniques to minimize current crowding.

Understanding SID mechanisms and their mitigation strategies is essential for ensuring the reliability and longevity of materials and components in various applications. By identifying and addressing these degradation processes, engineers can enhance the performance and safety of engineering systems.

4.2 Accelerated Life Testing (ALT) and Reliability Prediction

Accelerated Life Testing (ALT), also known as Highly Accelerated Life Testing (HALT), is a testing technique used to evaluate the reliability of a product or system under conditions that are more severe than normal operating conditions. The goal of ALT is to induce failures more quickly than would occur under normal operating conditions, allowing for the collection of reliability data in a shorter period of time.

ALT is typically performed by subjecting the product or system to a combination of elevated environmental stresses, such as high temperature, high humidity, and vibration. The level of stress is gradually increased until failures occur. The time to failure data is then used to estimate the reliability of the product or system under normal operating conditions.

ALT Planning

The planning of an ALT experiment is critical to its success. The following factors should be considered:

Test Objective: Clearly define the goals of the ALT experiment, such as estimating the reliability of a specific component or system.
Product Selection: Select products or systems that are representative of the actual operating environment.
Stress Levels: Determine the appropriate levels of environmental stress to induce failures within a reasonable timeframe.
Sample Size: Determine the number of products or systems to be tested to ensure statistical validity.
Data Collection: Plan for the collection of failure data, including the time to failure and the type of failure.

ALT Analysis

The analysis of ALT data involves the use of statistical models to estimate the reliability of the product or system under normal operating conditions. The following steps are typically involved:

Data Preprocessing: Clean and prepare the failure data

for analysis.
Model Selection: Choose an appropriate statistical model to fit the failure data, such as the Weibull or lognormal distribution.
Parameter Estimation: Estimate the parameters of the statistical model using the failure data.
Reliability Prediction: Calculate the reliability of the product or system under normal operating conditions using the estimated parameters.

Reliability Prediction

Reliability prediction is the process of estimating the reliability of a product or system based on its design and testing data. ALT data can be used to improve the accuracy of reliability predictions. The following steps are typically involved in reliability prediction:

Component Reliability: Estimate the reliability of individual components based on their failure rates.
System Reliability: Calculate the reliability of the system based on the reliability of its components and their interactions.
Failure Modes and Effects Analysis (FMEA): Identify potential failure modes and their impact on system reliability.
Reliability Allocation: Allocate reliability targets to different components or subsystems to ensure overall system reliability.

Benefits of ALT

ALT offers several benefits, including:

Time Savings: ALT can significantly reduce the time required to collect reliability data compared to traditional testing methods.
Cost Savings: ALT can reduce testing costs by allowing for the evaluation of multiple products or systems simultaneously.
Improved Reliability: ALT can help identify and mitigate potential reliability issues before a product is released to the market.
Increased Customer Confidence: ALT provides data to support reliability claims and enhance customer confidence in a product or system.

Limitations of ALT

ALT also has some limitations, including:

Assumptions: ALT assumes that the failure mechanisms under accelerated conditions are the same as those under normal operating conditions.
Extrapolation: Reliability predictions based on ALT data are extrapolations and may not always be accurate.
Cost: ALT can be expensive and may not be suitable for all products or systems.
Limited Sample Size: ALT experiments may involve a limited sample size, which can affect the accuracy of reliability predictions.

Conclusion

Accelerated Life Testing (ALT) is a valuable technique for evaluating the reliability of products and systems under accelerated conditions. ALT can provide significant benefits, including time and cost savings, improved reliability, and increased customer confidence. However, it is important to be aware of the limitations of ALT and to use it in conjunction with other reliability assessment techniques.

4.3 Reliability Metrics and Failure Analysis

It refers to the ability of a system to perform its intended function without failure over a specified period of time. In order to assess and improve the reliability of a system, it is important to have a clear understanding of the various reliability metrics and failure analysis techniques.

Reliability Metrics

Reliability metrics are quantitative measures that are used to describe the reliability of a system. These metrics can be used to compare different systems, track the reliability of a system over time, and identify areas for improvement. Some of the most common reliability metrics include:

Mean time to failure (MTTF): The average amount of time that a system is expected to operate before failing.

Mean time between failures (MTBF): The average amount of time between failures for a system that is repaired after each failure.
Failure rate: The number of failures that occur per unit of time.
Availability: The percentage of time that a system is operational.
Reliability: The probability that a system will not fail within a specified period of time.

Failure Analysis

Failure analysis is the process of identifying the root cause of a failure. This process can be complex and time-consuming, but it is essential for improving the reliability of a system. Failure analysis typically involves the following steps:

1. Data collection: Gather data about the failure, including the time of failure, the operating conditions, and any error messages.
2. Failure analysis: Analyze the data to identify the root cause of the failure. This may involve examining the failed component, reviewing the system logs, and interviewing the operators.
3. Corrective action: Develop and implement corrective actions to prevent the failure from recurring. This may involve redesigning the system, replacing the failed component, or changing the operating procedures.

Reliability Improvement

The goal of reliability improvement is to reduce the frequency and severity of failures. This can be achieved by implementing a variety of measures, including:

Design for reliability: Designing the system with reliability in mind. This involves using high-quality components, employing redundancy, and designing for ease of maintenance.
Quality control: Ensuring that the system is manufactured and assembled to high standards. This involves implementing quality control procedures and using qualified suppliers.
Preventive maintenance: Performing regular maintenance on the system to prevent failures from occurring. This involves cleaning, inspecting, and lubricating the system.
Failure analysis: Identifying the root cause of failures and implementing corrective actions to prevent them from recurring.

Conclusion

Reliability is a critical aspect of any system or component. By understanding the various reliability metrics and failure analysis techniques, it is possible to assess and improve the reliability of a system. This can lead to increased uptime, reduced maintenance costs, and improved customer satisfaction.

Chapter 5: Advanced Testing Techniques

5.1 High-Frequency Characterization: S-Parameters and Network Analysis

S-parameters, also known as scattering parameters, provide a comprehensive and efficient way to characterize the performance of high-frequency devices and circuits. Network analysis, utilizing S-parameters, enables engineers to accurately predict circuit behavior, analyze stability, and optimize designs.

S-Parameters: Definition and Measurement

S-parameters, often represented as a matrix, quantify the scattering behavior of a multi-port network, such as an amplifier, filter, or transmission line. Each element of the S-parameter matrix, denoted as Sij, represents the ratio of the reflected or transmitted signal at port i to the incident signal at port j. By measuring the S-parameters over a range of frequencies, engineers can fully characterize the device's behavior.

Vector network analyzers (VNAs) are specialized instruments used to measure S-parameters. VNAs generate signals at specific frequencies and inject them into the device under test, measuring the reflected and transmitted signals. The measured S-parameters provide insights into the device's impedance matching, gain, phase shift, and other critical characteristics.

Network Analysis Using S-Parameters

Once S-parameters are obtained, network analysis can be performed to analyze the behavior of circuits and systems. By connecting multiple S-parameter models together, engineers can simulate complex circuits and predict their overall performance. Network analysis techniques, such as cascading, series. parallel combinations, and feedback analysis, allow for the investigation of circuit stability, signal integrity, and noise characteristics.

Advantages of S-Parameter Characterization

S-parameter characterization offers numerous advantages for high-frequency circuit design:

Frequency-Domain Analysis: S-parameters provide a comprehensive view of a device's performance over a range of frequencies.
Multiple Port Characterization: S-parameters capture the interactions between multiple ports in a device, enabling the analysis of complex multi-port systems.
Stability Analysis: By examining the S-parameter matrix, engineers can assess the stability of circuits and identify potential oscillations.
Design Optimization: S-parameters provide valuable

insights for optimizing circuit designs to meet specific performance requirements.

Applications of S-Parameters and Network Analysis

S-parameter characterization and network analysis find applications in various fields, including:

Antenna Design: Optimizing antenna performance by matching impedance and minimizing reflections.
Microwave Circuits: Designing and analyzing high-frequency amplifiers, filters, and other microwave components.
RF Systems: Characterizing RF transceivers, power amplifiers, and other wireless devices.
Signal Integrity Analysis: Ensuring the integrity of high-speed signals in electronic circuits.
Automotive Electronics: Verifying the performance of electronic components in automotive applications.

Conclusion

S-parameters and network analysis are essential tools for characterizing and analyzing the behavior of high-frequency devices and circuits. By measuring and understanding S-parameters, engineers can accurately predict circuit performance, optimize designs, and ensure reliable operation in a wide range of applications. The comprehensive nature of S-parameter characterization makes it a powerful technique for advancing the design and analysis of high-frequency electronic systems.

5.2 Transient and Switching Measurements

In the realm of electrical engineering, transient and switching measurements play a crucial role in analyzing and understanding the behavior of electrical systems. Transient phenomena refer to short-lived, non-repetitive changes in electrical quantities, while switching measurements involve capturing the behavior of circuits during transitions between different states. Both types of measurements provide valuable insights into the dynamic performance of electrical systems.

Transient Measurements

Transient measurements aim to capture and analyze rapid changes in electrical signals that occur over a short period of time, typically ranging from nanoseconds to milliseconds. These transients can be caused by various events, such as switching operations, load variations, or lightning strikes. Understanding transient behavior is essential for ensuring the stability and reliability of electrical systems, as well as for preventing equipment damage.

Transient measurements can be performed using oscilloscopes, which are high-speed recording devices that display the time-varying behavior of electrical signals. Oscilloscopes allow for the precise measurement of transient amplitudes, durations, and other characteristics, providing detailed information about the transient event.

Common applications of transient measurements include:

Identifying and characterizing noise and interference in electrical systems
Evaluating the performance of power electronics circuits
Analyzing the response of electrical systems to sudden disturbances or faults
Measuring the switching characteristics of semiconductor devices

Switching Measurements

Switching measurements focus on capturing the behavior of electrical circuits during transitions between different states, such as when a switch is turned on or off or when a circuit element changes its state. These measurements provide insights into the dynamic behavior of circuits and can help identify potential issues or areas for improvement.

Switching measurements can be performed using a variety of techniques, including:

Capacitance-Voltage (C-V) Measurements: These measurements involve applying a varying voltage to a capacitor and measuring the resulting capacitance. C-V measurements provide information about the voltage dependence of the capacitor's characteristics, such as its dielectric constant.

Inductance-Current (L-I) Measurements: Similar to C-V measurements, L-I measurements involve applying a varying current to an inductor and measuring the resulting inductance. These measurements provide insights into the magnetic properties of the inductor, such as its core saturation.
Transfer Function Measurements: Transfer function measurements involve applying a sinusoidal input signal to a circuit and measuring the resulting output signal. The ratio of the output to the input signal provides the transfer function of the circuit, which describes its frequency response.

Common applications of switching measurements include:

Characterizing the switching characteristics of semiconductor devices, such as transistors and diodes
Measuring the frequency response of circuits and systems
Evaluating the stability and performance of feedback control systems

Significance of Transient and Switching Measurements

Transient and switching measurements are essential for gaining a comprehensive understanding of the dynamic behavior of electrical systems. By capturing and analyzing these phenomena, engineers can identify potential issues, optimize circuit performance, and ensure the reliability and safety of electrical systems.

Transient and switching measurements find applications in various fields, including:

Power electronics: Designing and analyzing power conversion circuits
Electronics: Characterizing semiconductor devices and designing electronic circuits
Communications: Analyzing signal integrity and optimizing communication systems
Control systems: Evaluating the stability and performance of control loops
Biomedical engineering: Measuring physiological signals and monitoring medical devices

Conclusion

Transient and switching measurements play a critical role in the analysis and design of electrical systems. By capturing and understanding these phenomena, engineers can gain valuable insights into the dynamic behavior of circuits and systems, ensuring their stability, reliability, and optimal performance.

5.3 Non-Destructive Testing and Probing Techniques

They provide a means to assess the structural integrity, material properties, and internal features of materials without causing any permanent damage. These techniques are widely employed in various industries, including manufacturing, construction, aerospace, automotive, and healthcare, to ensure the safety, reliability, and performance of materials and structures.

Principles of Non-Destructive Testing

NDT techniques utilize various physical phenomena to probe the internal structure and properties of materials. These phenomena include:

Ultrasound: High-frequency sound waves are introduced into the material, and the echoes and reflections are analyzed to determine the presence of defects, voids, or changes in material properties.
Electromagnetic waves: Eddy currents, magnetic fields, or radiographic waves are used to detect surface or subsurface flaws, measure material thickness, or identify variations in conductivity or permeability.
Thermal methods: Heat is applied to the material, and the temperature distribution or flow patterns are analyzed to reveal thermal properties, structural integrity, or the presence of defects.

Common NDT Techniques

Several NDT techniques are commonly used, each with its own advantages and limitations:

Ultrasonic testing (UT): Uses sound waves to detect internal defects and measure material thickness.
Eddy current testing (ET): Induces eddy currents in conductive materials to detect surface or near-surface

flaws.
Radiographic testing (RT): Uses X-rays or gamma rays to create images of internal structures and identify defects.
Magnetic particle testing (MT): Applies magnetic fields to ferromagnetic materials to detect surface-breaking defects.
Liquid penetrant testing (PT): Uses fluorescent or dye-penetrant liquids to reveal surface cracks or porosity.
Visual testing (VT): Involves direct observation of the material surface to identify visible defects or anomalies.

Applications of NDT

NDT techniques are used for a wide range of applications, including:

Defect detection: Identifying cracks, voids, inclusions, or other structural flaws that may compromise material integrity.
Material characterization: Determining material properties such as thickness, density, hardness, and elastic modulus.
Structural evaluation: Assessing the condition of bridges, buildings, pipelines, and other structures for safety and reliability.
Quality control: Ensuring that manufactured products meet specified quality standards and performance requirements.
Forensic investigations: Identifying the cause of failures or defects in materials and structures.

Probing Techniques

In addition to NDT techniques, probing techniques are also used to investigate the internal structure and properties of materials at a microscopic level. These techniques include:

Scanning probe microscopy (SPM): Uses a sharp tip to scan the surface of a material, providing atomic-scale images and information about surface morphology, topography, and material properties.
Transmission electron microscopy (TEM): Uses a high-energy electron beam to penetrate thin samples, providing detailed images of internal structures and crystal defects.
Scanning electron microscopy (SEM): Uses a focused electron beam to scan the surface of a material, providing high-resolution images and information about surface morphology, elemental composition, and crystal structure.

Advantages and Limitations of NDT and Probing Techniques

NDT and probing techniques offer several advantages:

Non-destructive: They do not damage the material being inspected.
Portable and versatile: They can be used in a variety of environments and on different types of materials.

Cost-effective: They are relatively inexpensive compared to destructive testing methods.

However, these techniques also have some limitations:

Limited depth of penetration: Some techniques may not be suitable for inspecting thick or opaque materials.
Sensitivity: The detection limit for defects or anomalies may vary depending on the technique and the material being inspected.
Operator dependence: The reliability of the results can depend on the skill and experience of the operator.

Conclusion

Non-destructive testing and probing techniques are powerful tools for characterizing materials and assessing their structural integrity and properties. By utilizing various physical phenomena, these techniques provide valuable information without causing any permanent damage to the material. The choice of technique depends on the specific application, material type, and desired level of detail. With continued advancements in NDT and probing technologies, the field of materials characterization continues to expand, enabling the development of safer, more reliable, and higher-performing materials and structures.

Chapter 6: Statistical Analysis and Data Interpretation

6.1 Data Acquisition and Handling

Data acquisition is the process of collecting and measuring information from the physical world. This information can be used to monitor and control systems, make decisions, and improve processes. Data acquisition systems are used in a wide variety of applications, including manufacturing, healthcare, transportation, and environmental monitoring.

The first step in data acquisition is to select the appropriate sensors. Sensors are devices that convert physical quantities, such as temperature, pressure, or flow, into electrical signals. The type of sensor that is used will depend on the specific application.

Once the sensors have been selected, they must be connected to a data acquisition system. Data acquisition systems can be either standalone devices or they can be integrated into a larger control system. Standalone data acquisition systems are typically used for simple applications, such as monitoring a single temperature sensor. Integrated data acquisition systems are used for more complex applications, such as controlling a manufacturing process.

Data acquisition systems typically include a number of features, such as:

Analog-to-digital converters (ADCs): ADCs convert analog signals from the sensors into digital signals that can be processed by the computer.
Digital-to-analog converters (DACs): DACs convert digital signals from the computer into analog signals that can be used to control actuators.
Input. output (I. O) ports: I. O ports allow the data acquisition system to communicate with other devices, such as sensors, actuators, and computers.
Software: Software is used to configure the data acquisition system and to process the data that is collected.

Data acquisition systems can be used to collect a wide variety of data. The type of data that is collected will depend on the specific application. Some of the most common types of data that are collected include:

Temperature
Pressure
Flow
Voltage
Current
Position
Speed
Acceleration

Data acquisition systems are essential for a wide variety of applications. By collecting and processing data from the physical world, data acquisition systems can help to improve efficiency, safety, and productivity.

Data Handling

Data handling is the process of managing and processing data. This includes tasks such as cleaning, transforming, and analyzing data. Data handling is an important step in the data analysis process, as it helps to ensure that the data is accurate and reliable.

There are a number of different data handling techniques that can be used, depending on the specific application. Some of the most common data handling techniques include:

Data cleaning: Data cleaning is the process of removing errors and inconsistencies from data. This can be done manually or using automated tools.
Data transformation: Data transformation is the process of converting data from one format to another. This can be done using a variety of techniques, such as rescaling, normalization, and binning.
Data analysis: Data analysis is the process of examining data to identify patterns and trends. This can be done using a variety of statistical and machine learning techniques.

Data handling is an important step in the data analysis process. By cleaning, transforming, and analyzing data, you can ensure that the data is accurate and reliable, and that you can draw valid conclusions from the data.

Conclusion

Data acquisition and handling are essential steps in the data analysis process. By collecting and processing data from the physical world, you can gain valuable insights into your processes and systems. This information can be used to improve efficiency, safety, and productivity.

6.2 Statistical Analysis and Hypothesis Testing

It is used in a wide variety of fields, including the social sciences, natural sciences, and business. Statistical analysis can be used to describe data, make predictions, and test hypotheses.

Hypothesis Testing

Hypothesis testing is a statistical method used to determine whether there is a statistically significant difference between two or more groups. A hypothesis is a statement about the population that is being tested. The null hypothesis (H0) is the statement that there is no difference between the groups. The alternative hypothesis (Ha) is the statement that there is a difference between the groups.

To test a hypothesis, we first collect data from the population. We then use statistical methods to calculate the probability of obtaining the data that we observed, assuming that the null hypothesis is true. If the probability is low, then we reject the null hypothesis and conclude that there is a statistically significant difference between the groups.

Statistical Significance

Statistical significance is a measure of the strength of the evidence against the null hypothesis. The p-value is the probability of obtaining the data that we observed, assuming that the null hypothesis is true. A p-value of 0. 05 or less is considered to be statistically significant.

Type I and Type II Errors

When we conduct a hypothesis test, we can make two types of errors:

Type I error: rejecting the null hypothesis when it is actually true.
Type II error: failing to reject the null hypothesis when it is actually false.

The probability of making a Type I error is controlled by the significance level, which is the maximum probability of rejecting the null hypothesis when it is actually true. The probability of making a Type II error is controlled by the sample size.

Sample Size

The sample size is the number of observations in a study. The larger the sample size, the more powerful the study will be. A more powerful study is less likely to make a Type II error.

Assumptions of Hypothesis Testing

Hypothesis testing assumes that the data are normally distributed and that the variances of the groups are equal. If these assumptions are not met, then the results of the hypothesis test may not be valid.

Statistical Analysis Software

Statistical analysis software can be used to perform a variety of statistical analyses, including hypothesis testing. There are a number of different statistical analysis software packages available, such as SPSS, SAS, and R.

Conclusion

Statistical analysis is a powerful tool that can be used to describe data, make predictions, and test hypotheses. Hypothesis testing is a statistical method used to determine whether there is a statistically significant difference between two or more groups. Statistical significance is a measure of the strength of the evidence against the null hypothesis. The probability of making a Type I error is controlled by the significance level, while the probability of making a Type II error is controlled by the sample size. Statistical analysis software can be used to perform a variety of statistical analyses, including hypothesis testing.

6.3 Statistical Process Control (SPC) and Failure Rate Analysis

Statistical process control (SPC) is a powerful tool that can be used to improve the quality of products and services. By monitoring and analyzing data from a process, SPC can help identify sources of variation and take steps to reduce or eliminate them. This can lead to improved product quality, reduced costs, and increased customer satisfaction.

Failure rate analysis (FRA) is a related technique that can be used to identify the root causes of failures. By analyzing data on failures, FRA can help identify design flaws, manufacturing defects, or other factors that may be contributing to the problem. This information can then be used to develop corrective actions to prevent future failures.

SPC and FRA are essential tools for any organization that is serious about improving quality. By using these techniques, organizations can identify and eliminate sources of variation and failure, leading to improved products and services and increased customer satisfaction.

SPC in Practice

SPC is a data-driven approach to quality control. It involves collecting data from a process, analyzing the data to identify sources of variation, and taking steps to reduce or eliminate those sources of variation.

One common SPC technique is the control chart. A control chart is a graph that plots data from a process over time. The control chart has two lines: a center line and an upper control limit (UCL) and a lower control limit (LCL). The center line represents the average value of the process, and the UCL and LCL represent the upper and lower limits of acceptable variation.

If the data points on the control chart fall within the UCL and LCL, then the process is said to be in control. This means that the process is stable and predictable, and there are no significant sources of variation.

However, if the data points on the control chart fall outside the UCL or LCL, then the process is said to be out of control. This means that there is a significant source of variation in the process, and steps need to be taken to identify and eliminate that source of variation.

SPC can be used to monitor and control any type of process. Some common applications of SPC include:

Manufacturing processes
Service processes
Business processes

FRA in Practice

FRA is a systematic approach to identifying the root causes of failures. It involves collecting data on failures, analyzing the data to identify patterns and trends, and developing corrective actions to prevent future failures.

One common FRA technique is the failure mode and effects analysis (FMEA). An FMEA is a table that lists all of the potential failure modes of a product or system, along with the effects of each failure mode and the likelihood of each failure mode occurring.

The FMEA can be used to identify the most critical failure modes and to develop corrective actions to prevent those failures from occurring.

FRA can be used to identify the root causes of failures in any type of product or system. Some common applications of FRA include:

Aerospace products
Automotive products
Medical devices
Software systems

Benefits of SPC and FRA

SPC and FRA are powerful tools that can help organizations improve quality and reduce costs. Some of the benefits of SPC and FRA include:

Improved product quality
Reduced costs
Increased customer satisfaction
Improved safety
Reduced risk

Conclusion

SPC and FRA are essential tools for any organization that is serious about improving quality. By using these techniques, organizations can identify and eliminate sources of variation and failure, leading to improved products and services and increased customer satisfaction.

Chapter 7: Transistor Design and Optimization

7.1 Design Considerations for Reliability and Performance

In the realm of software engineering, achieving reliability and performance are paramount concerns. These attributes are crucial for ensuring that software systems operate flawlessly, meet user expectations, and maintain a high level of availability. To attain these goals, meticulous attention must be paid to design considerations throughout the software development lifecycle. This section delves into the critical factors that influence reliability and performance and provides guidance on how to incorporate them into software design.

7. 1. 1 Reliability Considerations

Reliability refers to the ability of a software system to perform its intended functions without failure or unexpected behavior. Several design considerations contribute to enhancing reliability:

Modularity and Encapsulation: Decomposing a system into smaller, independent modules helps isolate potential errors and makes it easier to identify and fix them.

Encapsulating data and behavior within modules promotes information hiding, reducing the likelihood of unintended interactions.

Fault Tolerance: Incorporating mechanisms to handle and recover from failures is essential for ensuring system reliability. Techniques such as redundancy, error checking, and exception handling can prevent a single failure from cascading into a system-wide outage.

Testing and Validation: Rigorous testing at various levels (unit, integration, system, and acceptance) helps uncover and eliminate defects before the software is deployed. Validation ensures that the system meets its functional and non-functional requirements.

7. 1. 2 Performance Considerations

Performance encompasses the speed, responsiveness, and scalability of a software system. Key design considerations for optimizing performance include:

Data Structures and Algorithms: Choosing appropriate data structures and algorithms that efficiently store and manipulate data can significantly impact system performance. Factors to consider include time and space complexity, cache locality, and concurrency.

Caching and Indexing: Implementing caching mechanisms to store frequently accessed data in memory can reduce the time required to retrieve it. Indexing techniques, such as B-trees or hash tables, can accelerate data retrieval operations.

Concurrency and Parallelism: Exploiting concurrency and parallelism can improve performance by allowing multiple tasks or threads to execute simultaneously. However, it introduces challenges related to synchronization and communication between threads.

Scalability: Designing systems that can handle increasing workloads without significant performance degradation is crucial. Scalability can be achieved through techniques such as horizontal scaling (adding more servers) or vertical scaling (upgrading hardware).

7. 1. 3 Balancing Reliability and Performance

While reliability and performance are both important, there can be trade-offs between them. For example, implementing fault tolerance mechanisms can increase system complexity and potentially impact performance. It is essential to strike a balance between reliability and performance based on the specific requirements and constraints of the software system.

7. 1. 4 Design Patterns for Reliability and Performance

Design patterns provide reusable solutions to common software design problems. Several design patterns can be employed to enhance reliability and performance:

Observer Pattern: Decouples objects that need to be notified of changes, making it easier to add or remove observers without affecting the core logic.

Factory Method Pattern: Creates objects without specifying the exact class of the object that will be created, allowing for dynamic instantiation and reducing coupling.

Singleton Pattern: Ensures that a class has only one instance, promoting resource efficiency and simplifying access to shared resources.

7. 1. 5 Best Practices

In addition to the aforementioned considerations, adhering to best practices can further enhance reliability and performance:

Clean and Maintainable Code: Well-written, modular, and documented code reduces the likelihood of errors and facilitates maintenance and debugging.

Version Control and Code Review: Using version control systems and implementing code review processes promotes collaboration and helps identify potential issues early on.

Performance Monitoring and Profiling: Regularly monitoring system performance and using profiling tools can help identify performance bottlenecks and areas for improvement.

Continuous Integration and Deployment: Automating the build, testing, and deployment processes reduces the risk of errors and ensures consistent software quality.

Conclusion

Designing software systems for reliability and performance requires careful consideration of various factors throughout the software development lifecycle. By adhering to the principles outlined in this section, software engineers can create systems that are robust, efficient, and capable of meeting user expectations. Striking a balance between reliability and performance, leveraging design patterns, and adopting best practices are essential for achieving these goals.

7.2 Process Optimization and Variability Control

In the realm of manufacturing and process engineering, achieving optimal performance while minimizing variability is paramount to ensuring efficiency, quality, and customer satisfaction. Process optimization and variability control are two closely intertwined concepts that play a crucial role in achieving these goals.

Process Optimization

Process optimization aims to identify and adjust process

parameters to maximize desired outcomes while minimizing costs and waste. It involves a systematic approach that begins with defining process objectives and performance metrics. By understanding the process inputs and outputs, engineers can use statistical techniques and mathematical models to determine the optimal settings for process variables such as temperature, pressure, speed, and feed rates.

Optimization techniques range from simple trial-and-error methods to advanced statistical and mathematical algorithms. Six Sigma and Lean Manufacturing methodologies are widely used for process optimization, focusing on reducing defects, waste, and variation.

Variability Control

Variability control is essential for maintaining consistent product quality and preventing defects. It involves identifying and eliminating sources of variation in the process. Common sources of variation include raw material properties, equipment performance, operator skills, and environmental conditions.

To control variability, engineers use statistical process control (SPC) techniques. SPC involves collecting and analyzing data from the process to identify trends and patterns. Control charts are graphical tools used to monitor process performance and detect deviations from target values. When deviations occur, corrective actions are taken to bring the process back into control.

The Relationship between Process Optimization and Variability Control

Process optimization and variability control are closely related and mutually reinforcing. Optimized processes typically exhibit reduced variability, and controlled variability allows for more effective optimization. By iteratively applying these techniques, engineers can achieve a stable and efficient process that consistently produces high-quality products.

For instance, consider a manufacturing process for producing plastic parts. Process optimization might involve adjusting injection molding parameters to minimize cycle time and part weight. Variability control would focus on identifying and eliminating sources of variation in the molding process, such as temperature fluctuations or inconsistent material properties. By addressing variability, engineers can ensure that the optimized parameters continue to yield consistent results.

Benefits of Process Optimization and Variability Control

Implementing process optimization and variability control strategies offers numerous benefits for businesses, including:

Improved product quality and consistency
Reduced defects and waste

Increased production efficiency
Lower operating costs
Enhanced customer satisfaction

Conclusion

Process optimization and variability control are essential tools for manufacturers seeking to achieve operational excellence. By optimizing process parameters and minimizing variability, businesses can improve product quality, reduce costs, and enhance customer satisfaction. The iterative application of these techniques enables continuous improvement and sustained performance enhancements.

7.3 Design-for-Testability (DFT) Techniques

By incorporating DFT techniques into a design, engineers can improve the testability, diagnosability, and repairability of their circuits, leading to reduced testing costs, faster time-to-market, and higher product quality.

Benefits of DFT

DFT techniques offer several benefits, including:

Reduced testing costs: DFT techniques can reduce testing costs by making it easier to generate test vectors, perform fault simulations, and diagnose failures.
Faster time-to-market: DFT techniques can accelerate the development process by enabling faster testing and

debugging cycles.
Higher product quality: DFT techniques can improve product quality by ensuring that defects are detected and corrected early in the manufacturing process.
Increased diagnosability: DFT techniques can make it easier to diagnose failures and identify the root cause of problems.
Improved repairability: DFT techniques can make it easier to repair failed circuits by providing access to internal nodes and signals.

Types of DFT Techniques

There are various DFT techniques available, each with its own advantages and disadvantages. Some of the most common DFT techniques include:

Scan testing: Scan testing is a technique that involves inserting scan chains into a circuit. Scan chains are shift registers that allow internal nodes to be accessed and controlled externally. Scan testing enables easy generation of test vectors and fault simulations.
Boundary scan: Boundary scan is a technique that involves inserting a boundary scan register around the perimeter of a circuit. Boundary scan registers allow external access to internal signals and enable testing of interconnects and I. O devices.
Built-in self-test (BIST): BIST is a technique that involves incorporating self-testing logic into a circuit. BIST logic can automatically generate test vectors, perform fault simulations, and diagnose failures.

Design for debug: Design for debug techniques involve incorporating features into a circuit that make it easier to debug hardware and software issues. These features may include debug ports, trace buffers, and performance monitoring counters.

Choosing the Right DFT Technique

The choice of DFT technique depends on the specific requirements of the circuit. Factors to consider when choosing a DFT technique include:

Testability requirements: The level of testability required for the circuit.
Cost constraints: The cost of implementing the DFT technique.
Performance impact: The impact of the DFT technique on the circuit's performance.
Design complexity: The complexity of the circuit and the difficulty of implementing the DFT technique.

Conclusion

DFT techniques are essential for designing electronic circuits that are easy to test and debug. By incorporating DFT techniques into their designs, engineers can reduce testing costs, accelerate time-to-market, improve product quality, and increase diagnosability and repairability. The choice of DFT technique depends on the specific requirements of the circuit and should be carefully considered during the design process.

Chapter 8: Characterization of Emerging Transistor Technologies

8.1 Wide Bandgap Transistors: SiC, GaN, and Diamond

Wide bandgap (WBG) transistors are a class of semiconductor devices that have a wider bandgap than conventional silicon transistors. This wider bandgap gives WBG transistors several advantages over silicon transistors, including higher breakdown voltage, higher temperature operation, and faster switching speeds.

WBG transistors are made from materials such as silicon carbide (SiC), gallium nitride (GaN), and diamond. SiC and GaN are the most common WBG materials, and they are both used in a variety of power electronics applications. Diamond is a newer WBG material, and it has the potential to offer even higher performance than SiC and GaN.

8. 1. 1 Silicon Carbide (SiC)

Silicon carbide (SiC) is a compound semiconductor material that has a bandgap of 3. 26 eV. This wide bandgap gives SiC transistors several advantages over silicon transistors, including:

Higher breakdown voltage: SiC transistors can withstand higher voltages than silicon transistors. This makes them ideal for use in high-power applications, such as power supplies and motor drives.
Higher temperature operation: SiC transistors can operate at higher temperatures than silicon transistors. This makes them ideal for use in harsh environments, such as automotive and aerospace applications.
Faster switching speeds: SiC transistors have faster switching speeds than silicon transistors. This makes them ideal for use in high-frequency applications, such as telecommunications and radar systems.

SiC transistors are still relatively new, but they are rapidly gaining market share in a variety of applications. As the cost of SiC wafers continues to decline, SiC transistors are expected to become even more popular in the future.

8. 1. 2 Gallium Nitride (GaN)

Gallium nitride (GaN) is a compound semiconductor material that has a bandgap of 3. 4 eV. This wide bandgap gives GaN transistors several advantages over silicon transistors, including:

Higher breakdown voltage: GaN transistors can withstand higher voltages than silicon transistors. This makes them ideal for use in high-power applications, such as power supplies and motor drives.

Higher temperature operation: GaN transistors can operate at higher temperatures than silicon transistors. This makes them ideal for use in harsh environments, such as automotive and aerospace applications.
Faster switching speeds: GaN transistors have faster switching speeds than silicon transistors. This makes them ideal for use in high-frequency applications, such as telecommunications and radar systems.

GaN transistors are still relatively new, but they are rapidly gaining market share in a variety of applications. As the cost of GaN wafers continues to decline, GaN transistors are expected to become even more popular in the future.

8. 1. 3 Diamond

Diamond is a carbon-based semiconductor material that has a bandgap of 5. 5 eV. This wide bandgap gives diamond transistors several advantages over SiC and GaN transistors, including:

Even higher breakdown voltage: Diamond transistors can withstand even higher voltages than SiC and GaN transistors. This makes them ideal for use in ultra-high-power applications, such as power transmission and distribution systems.
Even higher temperature operation: Diamond transistors can operate at even higher temperatures than SiC and GaN transistors. This makes them ideal for use in extreme environments, such as space and nuclear

applications.
Even faster switching speeds: Diamond transistors have even faster switching speeds than SiC and GaN transistors. This makes them ideal for use in ultra-high-frequency applications, such as terahertz imaging and communications systems.

Diamond transistors are still in the early stages of development, but they have the potential to offer the highest performance of any WBG transistor. As the technology continues to mature, diamond transistors are expected to find applications in a wide range of industries, including power electronics, telecommunications, and aerospace.

8. 1. 4 Comparison of WBG Transistors

The following table compares the key characteristics of SiC, GaN, and diamond transistors:

Characteristic	SiC	GaN	Diamond
Bandgap (eV)	3. 26	3. 4	5. 5
Breakdown voltage (V)	>1000	>1000	>10,000
Temperature operation (°C)	>200	>200	>500
Switching speed (GHz)	>10	>10	>100

As you can see, diamond transistors offer the highest performance of any WBG transistor. However, they are also the most expensive and difficult to manufacture. SiC and GaN transistors offer a good compromise between performance and cost. They are less expensive and easier to manufacture than diamond transistors, but they still offer significant advantages over silicon transistors.

8. 1. 5 Applications of WBG Transistors

WBG transistors are used in a wide range of applications, including:

Power electronics: WBG transistors are ideal for use in power electronics applications, such as power supplies, motor drives, and inverters. They can handle high voltages and currents, and they have fast switching speeds.
Telecommunications: WBG transistors are used in a variety of telecommunications applications, such as base stations, repeaters, and amplifiers. They can handle high frequencies and they have low noise figures.
Aerospace: WBG transistors are used in a variety of aerospace applications, such as radar systems, satellite communications, and missile guidance systems. They can withstand harsh environments and they have high reliability.
Automotive: WBG transistors are used in a variety of automotive applications, such as engine control units, traction inverters, and battery chargers. They can handle high temperatures and they have fast switching speeds.

WBG transistors are a rapidly growing market, and they are expected to find applications in a wide range of industries in the future. As the cost of WBG wafers continues to decline, WBG transistors are expected to become even more popular.

8.2 2D Materials Transistors: Graphene and MoS2

Among these materials, graphene and molybdenum disulfide (MoS2) have attracted considerable attention for their potential applications in high-performance transistors. This section delves into the fundamental principles, device architectures, and performance characteristics of 2D materials transistors, with a focus on graphene and MoS2.

Graphene Transistors

Graphene, a single layer of carbon atoms arranged in a hexagonal lattice, possesses remarkable electrical properties. Its high carrier mobility, low contact resistance, and ballistic transport characteristics make it an ideal material for high-frequency and high-power transistors.

Device Architecture and Operation

Graphene transistors typically employ a field-effect transistor (FET) architecture, where the source and drain electrodes are connected to the graphene channel, and the gate electrode is separated from the channel by a dielectric layer. When a gate voltage is applied, it modulates the carrier concentration in the graphene channel, thereby controlling the flow of current between the source and drain.

Performance Characteristics

Graphene transistors exhibit excellent performance characteristics, including high on. off current ratios (exceeding 10^8), low subthreshold swing (below 60 mV. decade), and high transconductance. These properties enable graphene transistors to operate at high frequencies (up to THz) and with low power consumption. Additionally, graphene's mechanical flexibility and transparency make it suitable for flexible and transparent electronic devices.

MoS2 Transistors

MoS2, a layered transition metal dichalcogenide, is another promising 2D material for transistor applications. Unlike graphene, MoS2 is a semiconductor with a bandgap, which allows for efficient switching between the on and off states.

Device Architecture and Operation

MoS2 transistors also adopt an FET architecture, with source and drain electrodes connected to the MoS2 channel. The gate electrode is separated by a dielectric layer, and its voltage controls the carrier concentration in the MoS2 channel.

Performance Characteristics

MoS2 transistors exhibit a high on. off current ratio (exceeding 10^7), a subthreshold swing of around 70 mV. decade, and a moderate transconductance. Compared to graphene, MoS2 transistors have a lower carrier mobility, which limits their operating frequency. However, their semiconducting nature enables them to achieve high current drive and energy efficiency.

Advantages and Applications

2D materials transistors, such as graphene and MoS2, offer several advantages over conventional silicon-based transistors. Their high carrier mobility, low contact resistance, and unique electronic properties enable them to achieve superior performance characteristics. Additionally, their 2D nature allows for device miniaturization, flexibility, and transparency.

2D materials transistors have potential applications in a wide range of electronic devices, including high-frequency communication systems, low-power electronics, flexible displays, and sensors. Their unique properties make them particularly suitable for emerging technologies such as the Internet of Things (IoT), wearable devices, and artificial intelligence (AI).

Conclusion

Graphene and MoS2 are promising 2D materials for high-performance transistors. Their exceptional electrical properties, combined with their unique 2D nature, enable them to achieve superior performance characteristics and potential applications in next-generation electronic devices. Ongoing research efforts continue to explore the full potential of 2D materials transistors and push the boundaries of electronic device performance.

8.3 Organic Transistors

Organic materials are materials that are composed of carbon, hydrogen, and other elements. They are found in a wide variety of natural and synthetic materials, including plants, animals, and plastics.

Organic transistors have a number of advantages over traditional inorganic transistors. They are lightweight, flexible, and inexpensive to produce. They are also transparent, which makes them ideal for use in applications such as displays and sensors.

Basic Operation

Organic transistors operate on the same basic principles as inorganic transistors. They have three terminals: a source, a drain, and a gate. When a voltage is applied to the gate, it controls the flow of current between the source and the drain.

The key difference between organic transistors and inorganic transistors is the material that is used to make the channel. In inorganic transistors, the channel is made from a semiconductor material, such as silicon. In organic transistors, the channel is made from an organic material, such as a polymer or a small molecule.

Types of Organic Transistors

There are two main types of organic transistors:

Thin-film transistors (TFTs): TFTs are made by depositing a thin film of organic material onto a substrate. TFTs are the most common type of organic transistor and are used in a wide variety of applications.
Single-crystal transistors (SCOTs): SCOTs are made by growing a single crystal of organic material on a substrate. SCOTs have higher performance than TFTs, but they are more difficult to manufacture.

Applications

Organic transistors have a wide range of potential applications, including:

Displays: Organic transistors can be used to make flexible and transparent displays.
Sensors: Organic transistors can be used to make sensors for a variety of applications, such as chemical and biological sensing.
Radio frequency identification (RFID) tags: Organic transistors can be used to make RFID tags that are small, lightweight, and inexpensive.

Transistors for logic circuits: Organic transistors can be used to make logic circuits that are lightweight, flexible, and inexpensive.

Challenges

There are a number of challenges that need to be overcome before organic transistors can be widely adopted. These challenges include:

Stability: Organic materials are not as stable as inorganic materials. This can lead to problems with the performance and reliability of organic transistors.
Performance: The performance of organic transistors is not as good as the performance of inorganic transistors. This is due to a number of factors, including the lower mobility of organic materials.
Manufacturing: Organic transistors are more difficult to manufacture than inorganic transistors. This is due to the fact that organic materials are more sensitive to processing conditions.

Conclusion

Organic transistors have a number of potential advantages over traditional inorganic transistors. However, there are a number of challenges that need to be overcome before organic transistors can be widely adopted. Despite these challenges, organic transistors are a promising new technology with the potential to revolutionize a wide range of applications.

Chapter 9: Automation and Data Management

9.1 Automated Test Systems (ATS)

Definition and Overview

Automated Test Systems (ATS) are comprehensive software and hardware solutions designed to automate the testing process for electronic devices, components, and systems. ATSs leverage advanced technologies and techniques to execute a wide range of tests with precision, efficiency, and repeatability, enabling industries to ensure the reliability and quality of their products.

Components of an ATS

An ATS typically consists of several key components:

Test Executive Software: Coordinates the entire testing process, managing test sequences, data collection, and reporting.
Test Station Hardware: Provides the physical interface between the test system and the device under test (DUT).
Test Instruments: Measure and analyze the DUT's electrical or functional parameters.

Test Fixtures: Hold and connect the DUT to the test station hardware, ensuring proper contact and signal integrity.
Data Acquisition and Analysis Tools: Collect and process test data for analysis and evaluation.

Types of ATSs

Depending on the industry and application requirements, ATSs can be classified into various types:

In-Circuit Testers (ICTs): Verify the electrical connectivity and component values of PCBs.
Functional Testers: Test the overall functionality of electronic assemblies by simulating real-world operating conditions.
Boundary-Scan Testers (BSTs): Access and test specific nodes within ICs and complex devices.
Automatic Optical Inspection (AOI) Systems: Inspect PCBs for defects and solder joint quality using optical imaging.
X-ray Inspection Systems: Detect internal defects and component placement accuracy using X-ray technology.

Benefits of Using ATSs

ATSs offer numerous advantages over manual testing methods:

Increased Efficiency: Automate repetitive and time-consuming testing tasks, reducing labor costs and

improving throughput.
Improved Accuracy and Repeatability: Eliminate human error and ensure consistent test results, leading to higher reliability.
Comprehensive Coverage: Execute a wide range of tests to thoroughly verify device performance.
Reduced Test Time: Streamline the testing process, resulting in faster time-to-market.
Enhanced Quality Control: Identify and isolate defects early on, preventing faulty products from reaching customers.

Applications of ATSs

ATSs are extensively used in various industries, including:

Electronics Manufacturing: Testing PCBs, components, and electronic assemblies.
Automotive Industry: Verifying the functionality of automotive electronic systems.
Medical Device Development: Testing medical devices for safety, efficacy, and compliance.
Semiconductor Manufacturing: Characterizing and testing ICs and semiconductor devices.
Aerospace and Defense: Evaluating the reliability and performance of military and aerospace systems.

Future Trends in ATS

The future of ATSs is poised for advancements in:

Artificial Intelligence (AI): Incorporating AI algorithms for test planning, data analysis, and anomaly detection.
Internet of Things (IoT): Connecting ATSs to the IoT for remote monitoring and data sharing.
Virtual and Augmented Reality (VR. AR): Enhancing the user experience and training through virtual test environments.
Cloud Computing: Leveraging cloud platforms for data storage, analysis, and collaboration.
Cybersecurity: Addressing cybersecurity concerns associated with the increasing connectivity of ATSs.

Conclusion

Automated Test Systems (ATSs) are essential tools in the modern electronics industry, enabling the efficient and reliable testing of electronic devices, components, and systems. By automating the testing process, ATSs increase productivity, improve accuracy, and ensure product quality. As technology continues to evolve, ATSs will continue to advance, incorporating cutting-edge techniques to meet the ever-growing demands of the electronics industry.

9.2 Data Acquisition, Storage, and Analysis Software

Data acquisition, storage, and analysis software play a crucial role in the modern scientific research process. These tools enable researchers to efficiently collect, manage, and analyze vast amounts of data generated from experiments, simulations, and other sources. The availability of powerful and user-friendly software has significantly enhanced the productivity and accuracy of scientific investigations.

Data Acquisition

Data acquisition software is responsible for capturing and converting raw data from various sources, such as sensors, instruments, and laboratory equipment. It serves as an interface between the data source and the computer, allowing researchers to control data collection parameters and ensure data integrity. Modern data acquisition software offers advanced features like real-time monitoring, data validation, and automated error handling. These capabilities help researchers optimize the data acquisition process and minimize the risk of data loss or errors.

Data Storage

Once data is acquired, it needs to be stored securely and efficiently for future analysis and reference. Data storage software provides a structured framework for organizing and managing large datasets. These software solutions offer features such as hierarchical data organization, data compression, and encryption to ensure data security and integrity. Advanced data storage software also supports distributed storage systems, enabling researchers to store and access data across multiple locations and computing platforms.

Data Analysis

Data analysis software is the cornerstone of scientific research, allowing researchers to extract meaningful insights from raw data. These tools provide a comprehensive suite of statistical, graphical, and machine learning algorithms for data exploration, hypothesis testing, and predictive modeling. Researchers can use data analysis software to identify trends, correlations, and patterns in their data, helping them formulate conclusions and advance their research. Modern data analysis software often integrates with data visualization tools, enabling researchers to create interactive plots and visualizations that facilitate data understanding and communication.

Integration and Automation

Data acquisition, storage, and analysis software often work together as an integrated system, streamlining the research workflow and minimizing manual errors. Researchers can use software suites that combine all these functions into a single platform, providing a seamless experience from data collection to analysis. Automation features, such as scheduled data backups and automated data analysis routines, further enhance the efficiency and reliability of the research process.

Importance of Data Management

Effective data management is essential for successful scientific research. Data acquisition, storage, and analysis software provide the foundation for robust data management practices. By ensuring data integrity, security, and accessibility, these tools empower researchers to conduct rigorous and reproducible research. Moreover, well-managed data enables researchers to share and collaborate with others, fostering scientific progress and innovation.

Challenges and Considerations

While data acquisition, storage, and analysis software offer numerous benefits, researchers should be aware of potential challenges and considerations. These include:

Data Volume and Complexity: Modern research often generates massive and complex datasets, requiring specialized software and hardware solutions for efficient management and analysis.
Data Security and Privacy: Protecting sensitive data from unauthorized access and misuse is paramount. Researchers must implement appropriate security measures and adhere to ethical guidelines for data handling.
Software Compatibility: Researchers may need to work with different software tools and formats during their research. Ensuring compatibility and interoperability between these tools is crucial for seamless data exchange and analysis.
User Training and Support: Researchers require proper

training and technical support to effectively utilize data management software. Software providers should offer comprehensive documentation, tutorials, and dedicated support channels.

Conclusion

Data acquisition, storage, and analysis software are indispensable tools for modern scientific research. These tools empower researchers to efficiently collect, manage, and analyze vast amounts of data, leading to advancements in knowledge and innovation. By understanding the capabilities and considerations associated with these software solutions, researchers can optimize their data management practices and unlock the full potential of their research endeavors.

9.3 Data Visualization and Reporting

Data visualization is the graphical representation of data. It helps to make data more accessible and easier to understand. Data visualization can be used for a variety of purposes, such as:

Exploring data to identify patterns and trends
Communicating data to others in a clear and concise way
Making decisions based on data

There are many different types of data visualizations, including charts, graphs, and maps. The type of visualization that you choose will depend on the data you have and the purpose of your visualization.

Here are some of the most common types of data visualizations:

Bar charts: Bar charts are used to compare the values of different categories. They are created by drawing bars that are proportional to the values of the categories.
Line charts: Line charts are used to show how a value changes over time. They are created by drawing a line that connects the data points.
Pie charts: Pie charts are used to show the proportions of a whole. They are created by dividing a circle into slices that are proportional to the values of the categories.
Scatterplots: Scatterplots are used to show the relationship between two variables. They are created by plotting the data points on a graph and drawing a line that shows the trend of the data.

Data visualization can be a powerful tool for understanding and communicating data. By using the right type of visualization, you can make your data more accessible and easier to understand for others.

Reporting

Reporting is the process of presenting data in a clear and concise way. Reports can be used for a variety of purposes, such as:

Summarizing data for decision-making
Communicating data to others
Documenting data for future reference

There are many different types of reports, including:

Narrative reports: Narrative reports are written in a narrative style and tell a story about the data.
Analytical reports: Analytical reports provide an analysis of the data and identify patterns and trends.
Informational reports: Informational reports provide a summary of the data without any analysis.

The type of report that you choose will depend on the purpose of your report and the audience you are writing for.

Here are some tips for writing effective reports:

Start with a strong introduction.
Organize the data in a logical way. The data should be organized in a way that makes it easy to understand.
Use clear and concise language. The language should be easy to understand and free of jargon.
Use visuals to support your data. Visuals can help to make the data more accessible and easier to understand.
Proofread your report carefully. Make sure that there are no errors in grammar or spelling.

Reporting can be a valuable tool for communicating data. By using the right type of report and following the tips above, you can create reports that are clear, concise, and effective.

Conclusion

Data visualization and reporting are essential skills for anyone who works with data. Data visualization can help you to understand data and communicate it to others in a clear and concise way. Reporting can help you to summarize data for decision-making, communicate data to others, and document data for future reference.

By mastering the skills of data visualization and reporting, you can become a more effective data analyst and communicator.

Chapter 10: Application of Transistor Characterization

10.1 Reliability Assessment for Electronic Devices

Electronic devices play a crucial role in our modern world, from powering our smartphones and laptops to controlling critical infrastructure like medical equipment and transportation systems. Ensuring the reliability of these devices is essential for both safety and performance. Reliability assessment provides a systematic approach to evaluate the likelihood of a device failing over time.

10. 1. 1 Importance of Reliability Assessment

Reliability assessment helps manufacturers and users to:

Predict device failures: Identify potential failure modes and estimate the probability of failure.
Optimize designs: Improve device designs to reduce failure rates and extend operating lifetimes.
Establish maintenance schedules: Determine the frequency of maintenance interventions to prevent failures.
Ensure safety and compliance: Meet regulatory requirements and industry standards for device

reliability.

10. 1. 2 Types of Reliability Assessment Methods

Various methods are employed for reliability assessment, including:

Accelerated Life Testing (ALT): Exposes devices to accelerated conditions (e. g. , high temperature, voltage) to rapidly induce failures and estimate failure rates.
Field Data Analysis: Collects failure data from devices in real-world use to assess actual reliability in the operating environment.
Physics-of-Failure Analysis: Identifies the physical mechanisms (e. g. , wear-out, thermal fatigue) that lead to device failures.
Reliability Modeling: Uses statistical models to predict failure rates based on device characteristics and operating conditions.

10. 1. 3 Key Parameters for Reliability Assessment

Reliability assessment focuses on quantifying the following parameters:

Mean Time to Failure (MTTF): Average time until a device fails.
Failure Rate: Probability of failure per unit of time or usage.
Reliability Function: Probability of a device surviving a given operating time without failure.

10. 1. 4 Steps Involved in Reliability Assessment

Reliability assessment typically involves the following steps:

1. Define device specifications: Determine the operating conditions and performance requirements.
2. Identify failure modes: Analyze potential causes of device failure.
3. Select reliability assessment method: Choose the appropriate method based on device characteristics and available resources.
4. Conduct data collection: Gather failure data through testing or field observation.
5. Analyze data: Estimate reliability parameters and identify failure mechanisms.
6. Make recommendations: Propose design improvements, maintenance schedules, or other measures to enhance reliability.

10. 1. 5 Challenges in Reliability Assessment

Reliability assessment faces several challenges, including:

Device complexity: Modern electronic devices are highly complex, making failure analysis difficult.
Limited failure data: In some cases, insufficient failure data is available to perform accurate reliability estimates.
Environmental factors: Operating conditions and

environmental stressors can significantly impact device reliability.
Continuous innovation: Rapid technological advancements introduce new challenges and require ongoing reliability assessments.

Despite these challenges, reliability assessment remains an essential tool for ensuring the safe and reliable operation of electronic devices in various applications. By understanding the principles and methods of reliability assessment, engineers can design, manufacture, and maintain electronic devices with confidence and minimize the risk of failures.

10.2 Failure Analysis and Root Cause Investigation

Failure analysis and root cause investigation are critical processes for understanding why failures occur and preventing them from happening again. Failure analysis is the process of examining a failed component or system to determine the cause of the failure. Root cause investigation is the process of identifying the underlying causes of a failure and developing corrective actions to prevent it from happening again.

Failure Analysis

The first step in failure analysis is to collect data about the failure. This data can include:

The failed component or system
The operating conditions at the time of the failure
The maintenance history of the component or system
The results of any previous failure analyses

Once the data has been collected, it is analyzed to determine the cause of the failure. The analysis may involve:

Visual inspection
Microscopy
Chemical analysis
Mechanical testing
Electrical testing

The results of the analysis are used to develop a failure mechanism. The failure mechanism is a description of the sequence of events that led to the failure.

Root Cause Investigation

The purpose of root cause investigation is to identify the underlying causes of a failure and develop corrective actions to prevent it from happening again. The root cause investigation process typically involves the following steps:

1. Define the problem. The first step is to define the problem that occurred. This involves gathering information about the failure and understanding the impact it had on the organization.

2. Identify the root cause. The next step is to identify the root cause of the failure. This involves analyzing the data collected during the failure analysis and identifying the underlying causes of the failure.
3. Develop corrective actions. Once the root cause has been identified, corrective actions can be developed to prevent the failure from happening again. These actions may include changes to the design of the component or system, changes to the operating procedures, or changes to the maintenance schedule.
4. Implement corrective actions. The final step is to implement the corrective actions. This involves making the necessary changes to the component or system and monitoring the results to ensure that the failure does not happen again.

Benefits of Failure Analysis and Root Cause Investigation

Failure analysis and root cause investigation can provide a number of benefits for organizations, including:

Reduced downtime. By understanding the causes of failures, organizations can take steps to prevent them from happening again, which can reduce downtime and improve productivity.
Improved safety. Failures can lead to safety hazards, so by understanding the causes of failures, organizations can take steps to prevent them from happening again and improve safety.
Reduced costs. Failures can be costly, so by understanding the causes of failures, organizations can

take steps to prevent them from happening again and reduce costs.
Improved quality. Failures can lead to poor quality products or services, so by understanding the causes of failures, organizations can take steps to prevent them from happening again and improve quality.

Conclusion

Failure analysis and root cause investigation are critical processes for understanding why failures occur and preventing them from happening again. By conducting thorough failure analyses and root cause investigations, organizations can reduce downtime, improve safety, reduce costs, and improve quality.

10.3 Process Control and Improvement

Process control is the use of control loops to maintain a process at a desired setpoint. Control loops consist of three main components: a sensor, a controller, and an actuator. The sensor measures the process variable and sends this information to the controller. The controller compares the measured process variable to the setpoint and calculates an error signal. The error signal is then sent to the actuator, which makes adjustments to the process to reduce the error.

There are many different types of control loops, but the most common is the proportional-integral-derivative (PID) controller. PID controllers use a combination of proportional, integral, and derivative action to control the process variable. Proportional action adjusts the output of the controller in proportion to the error signal. Integral action eliminates steady-state error by adjusting the output of the controller based on the integral of the error signal. Derivative action anticipates changes in the process variable and adjusts the output of the controller accordingly.

Process control is used in a wide variety of industries, including manufacturing, pharmaceuticals, and food processing. By maintaining a process at a desired setpoint, process control can improve product quality, reduce costs, and increase efficiency.

Process Improvement

Process improvement is the systematic approach to making processes more efficient and effective. Process improvement can be applied to any type of process, from manufacturing processes to business processes. The goal of process improvement is to identify and eliminate waste and inefficiency, and to improve the overall performance of the process.

There are many different process improvement methodologies, but the most common is the DMAIC (Define, Measure, Analyze, Improve, Control) methodology. DMAIC is a five-step process that involves defining the problem, measuring the current performance of the process, analyzing the data to identify the root causes of the problem, implementing improvements to address the root causes, and controlling the process to ensure that the improvements are sustained.

Process improvement can bring a number of benefits, including reduced costs, improved quality, increased efficiency, and improved customer satisfaction. By systematically identifying and eliminating waste and inefficiency, process improvement can help organizations achieve their goals and improve their overall performance.

Process Control and Improvement in Practice

Process control and process improvement are two essential tools for improving the performance of any organization. By using these tools, organizations can identify and eliminate waste and inefficiency, improve product quality, reduce costs, and increase efficiency.

Here are some examples of how process control and improvement have been used in practice:

Manufacturing: A manufacturing company used process control to improve the quality of its products. The company installed sensors on its production line to measure the dimensions of its products. The data from the sensors was then used to adjust the production process in real time, ensuring that the products met the desired specifications.
Pharmaceuticals: A pharmaceutical company used process

improvement to reduce the cost of its products. The company used the DMAIC methodology to identify and eliminate waste and inefficiency in its production process. As a result, the company was able to reduce the cost of its products by 15%.
Food processing: A food processing company used process control to improve the safety of its products. The company installed sensors on its production line to detect contamination. The data from the sensors was then used to adjust the production process in real time, ensuring that the products were safe for consumption.

These are just a few examples of how process control and improvement can be used to improve the performance of organizations. By using these tools, organizations can identify and eliminate waste and inefficiency, improve product quality, reduce costs, and increase efficiency.

Conclusion

Process control and process improvement are essential tools for improving the performance of any organization. By using these tools, organizations can identify and eliminate waste and inefficiency, improve product quality, reduce costs, and increase efficiency.

Chapter 11: Emerging Trends in Transistor Testing

11.1 Advancements in Testing Equipment and Techniques

The field of materials testing has undergone significant advancements in recent years, particularly in the development of testing equipment and techniques. These advancements have enabled researchers and engineers to characterize materials with greater accuracy, precision, and efficiency, leading to improved product design and performance.

One of the most notable advancements has been the development of non-destructive testing (NDT) techniques. NDT methods allow for the evaluation of materials without causing any damage to the specimen, making them ideal for inspecting critical components and structures in service. Common NDT techniques include ultrasonic testing, radiography, and eddy current testing. Ultrasonic testing uses high-frequency sound waves to detect internal defects and flaws, while radiography employs X-rays or gamma rays to visualize the internal structure of materials. Eddy current testing, on the other hand, uses electromagnetic induction to detect surface defects and cracks.

Another significant advancement has been the development of advanced microscopy techniques. Electron microscopy, such as scanning electron microscopy (SEM) and transmission electron microscopy (TEM), provides high-resolution images of materials at the nanoscale. These techniques allow researchers to study the microstructure of materials, including grain size, morphology, and defects. By understanding the microstructure, engineers can tailor the properties of materials for specific applications.

Furthermore, the development of automated testing systems has greatly improved the efficiency and accuracy of materials testing. These systems use computer-controlled equipment to perform tests according to predefined protocols. Automated testing systems can reduce human error, increase throughput, and provide real-time data analysis. They are particularly beneficial for repetitive and high-volume testing applications.

In addition to these major advancements, numerous other improvements have been made in testing equipment and techniques. These include the development of high-temperature and low-temperature testing chambers, specialized fixtures for testing specific materials, and advanced data acquisition and analysis software. These advancements have collectively contributed to the enhanced capabilities and reliability of materials testing.

Impact of Advancements on Materials Testing

The advancements in testing equipment and techniques have had a profound impact on the field of materials testing. These advancements have:

Increased accuracy and precision: Advanced equipment and techniques enable researchers and engineers to obtain more accurate and precise measurements of material properties. This leads to more reliable and trustworthy data for product design and development.

Expanded testing capabilities: New testing techniques, such as NDT and advanced microscopy, have expanded the range of materials that can be tested and the properties that can be characterized. This allows researchers and engineers to explore new materials and applications.

Improved efficiency: Automated testing systems and other advancements have significantly improved the efficiency of materials testing. This enables faster and more cost-effective testing, which is essential for high-volume production and quality control.

Enhanced safety: NDT techniques, in particular, have improved the safety of materials testing by eliminating the need for destructive testing. This is especially important for testing critical components and structures in service.

Future Trends in Testing Equipment and Techniques

The field of materials testing is continuously evolving, and new advancements in testing equipment and techniques are emerging. Some of the key trends for the future include:

Integration of advanced sensing technologies: The integration of advanced sensing technologies, such as fiber optics and microelectronics, will enable the development of more sensitive and versatile testing equipment. This will allow for the detection and characterization of materials at even smaller scales.

Development of artificial intelligence (AI) and machine learning (ML) algorithms: AI and ML algorithms will play an increasingly important role in materials testing. These algorithms can be used to analyze large amounts of testing data and identify patterns that may not be apparent to human inspectors. This will lead to more accurate and reliable testing results.

Increased use of non-destructive testing (NDT) techniques: NDT techniques will continue to gain popularity due to their safety and efficiency advantages. New NDT techniques are being developed to address the challenges of testing complex materials and structures.

These future trends will further enhance the capabilities of materials testing and contribute to the development of better and more reliable products.

11.2 Big Data Analytics and Machine Learning in Characterization

This data encompasses electronic health records (EHRs), medical images, genomic data, and various other sources. Traditional data analysis methods are often insufficient to handle such large and diverse datasets, which has led to the adoption of advanced techniques such as big data analytics and machine learning. These techniques enable researchers and clinicians to extract meaningful insights from vast amounts of data, leading to improved patient care, drug discovery, and clinical decision-making.

Big Data Analytics

Big data analytics involves the analysis of large, complex datasets that are too voluminous and complex for traditional data processing software. The key characteristics of big data are often summarized by the "5 Vs":

Volume: The size of the data is massive, often exceeding terabytes or even petabytes.
Variety: The data comes in various forms, including structured (e. g. , EHRs), unstructured (e. g. , medical notes), and semi-structured (e. g. , genomic data).
Velocity: The data is constantly generated and updated at a rapid pace.
Veracity: The data may contain errors, inconsistencies, and missing values.

Value: The data has the potential to provide valuable insights and improve decision-making.

Big data analytics techniques enable researchers to uncover hidden patterns, correlations, and trends in the data. These insights can be used to identify risk factors for diseases, optimize treatment plans, and predict patient outcomes.

Machine Learning

Machine learning is a subset of artificial intelligence that allows computers to learn from data without explicit programming. Machine learning algorithms are trained on large datasets and can then make predictions or classifications on new data. The most common types of machine learning algorithms include:

Supervised learning: The algorithm is trained on labeled data, where the input data is paired with the desired output. The algorithm learns the relationship between the input and output and can then predict the output for new, unseen data.
Unsupervised learning: The algorithm is trained on unlabeled data, where the input data does not have an associated output. The algorithm learns to find patterns and structures in the data without guidance.
Reinforcement learning: The algorithm learns by interacting with its environment and receiving feedback on its actions. The algorithm adjusts its behavior over time to maximize its rewards and minimize its

punishments.

Applications in Characterization

Big data analytics and machine learning have numerous applications in characterization, including:

Patient stratification: Identifying subgroups of patients with similar characteristics, such as disease severity, treatment response, or risk of adverse events.
Predictive modeling: Predicting patient outcomes, such as disease progression, response to treatment, or risk of hospitalization.
Disease subtyping: Identifying different subtypes of diseases based on their molecular profiles or clinical characteristics.
Drug discovery: Identifying potential drug targets and optimizing drug development processes.
Clinical decision support: Providing clinicians with real-time guidance on diagnosis, treatment, and patient management.

Challenges and Limitations

While big data analytics and machine learning offer great potential for improving characterization, there are also challenges and limitations associated with these techniques:

Data quality and availability: The quality and availability of data can vary widely, which can impact the accuracy

and reliability of the analysis.
Interpretability: Machine learning models can be complex and difficult to interpret, making it challenging to understand the underlying decision-making process.
Bias and fairness: Machine learning algorithms can be biased if the training data is not representative of the population of interest. This can lead to unfair or discriminatory outcomes.
Ethical considerations: The use of big data analytics and machine learning raises ethical concerns related to privacy, data security, and algorithmic bias.

Conclusion

Big data analytics and machine learning are powerful tools that have the potential to revolutionize characterization. These techniques enable researchers and clinicians to extract meaningful insights from vast amounts of data, leading to improved patient care, drug discovery, and clinical decision-making. However, it is important to be aware of the challenges and limitations associated with these techniques and to use them responsibly and ethically.

11.3 Future Directions in Transistor Testing and Reliability

The relentless pursuit of smaller, faster, and more power-efficient transistors has driven the semiconductor industry to its limits. As transistor dimensions continue to shrink, the challenges associated with testing and ensuring their reliability become increasingly daunting. This section explores some of the key future directions in transistor testing and reliability that will be crucial to enabling continued progress in the semiconductor industry.

Advanced Testing Techniques:

Traditional transistor testing methods are no longer sufficient to characterize the complex behavior of modern transistors. Advanced testing techniques, such as high-frequency testing, stress testing, and noise analysis, will be essential for accurately assessing transistor performance and reliability. These techniques can provide insights into transistor behavior under various operating conditions, enabling designers to optimize transistor design and predict device lifetime.

Artificial Intelligence (AI) for Testing and Reliability:

AI is revolutionizing many aspects of semiconductor manufacturing and testing. AI algorithms can be trained on vast datasets of transistor measurements to identify patterns and anomalies that may not be easily detectable by traditional methods. AI-powered testing systems can automate test procedures, improve accuracy, and reduce test times. Additionally, AI can be used to predict transistor reliability based on historical data and environmental factors, enabling proactive maintenance and failure prevention.

Reliability Enhancement Techniques:

As transistor dimensions shrink, the susceptibility to defects and reliability concerns increases. Novel reliability enhancement techniques will be required to ensure the long-term performance and stability of transistors. These techniques may include advanced packaging materials, stress-relieving structures, and self-healing mechanisms. Researchers are exploring innovative approaches to mitigate the effects of electromigration, thermal fatigue, and other reliability-limiting factors.

Integration with Process Monitoring:

Effective transistor testing and reliability assessment require close integration with process monitoring. In-line process control systems can provide real-time feedback on transistor characteristics during manufacturing, enabling adjustments to the process to optimize device performance and reliability. This integration will be critical for achieving consistent and high-quality transistor production.

Advanced Characterization Tools:

The development of advanced characterization tools is essential for understanding the fundamental mechanisms that govern transistor behavior and reliability. These tools include scanning probe microscopy, transmission electron microscopy, and X-ray diffraction. By providing detailed insights into transistor structure, material properties, and defect distribution, these tools enable researchers to identify and address reliability concerns at the atomic level.

Predictive Modeling:

Predictive modeling is a powerful tool for forecasting transistor reliability. By combining experimental data with physics-based models, researchers can develop predictive models that can estimate device lifetime and identify potential failure mechanisms. These models can guide design decisions and enable proactive measures to mitigate reliability risks. The future directions outlined in this section provide a glimpse into the exciting challenges and opportunities that lie ahead for transistor testing and reliability. By harnessing the power of advanced techniques, AI, reliability enhancement, process integration, characterization tools, and predictive modeling, researchers and engineers will pave the way for the next generation of highly reliable and efficient transistors that will power the technologies of the future.

Chapter 12: Case Studies and Examples

12.1 Examples of Transistor Characterization in Different Applications

Transistors are essential components in modern electronic devices, and their characterization is crucial to ensure their proper operation and reliability. Different applications impose specific requirements on transistors, and the characterization techniques used must be tailored accordingly. In this section, we will discuss some examples of transistor characterization in different applications, highlighting the specific parameters and techniques involved.

12. 1. 1 Power Electronics

Transistors used in power electronics applications are typically characterized for their ability to handle high currents and voltages. The following parameters are of primary interest:

Breakdown voltage (BV): The maximum voltage that the transistor can withstand without breaking down.
On-state resistance (RDS(on)): The resistance of the transistor when it is turned on.
Off-state leakage current (IDSS): The current that

flows through the transistor when it is turned off.
Gate-source threshold voltage (VGS(th)): The minimum voltage that must be applied to the gate to turn on the transistor.
Switching time (t): The time it takes for the transistor to switch from on to off or vice versa.

The characterization of power transistors typically involves measuring these parameters under various operating conditions, such as different temperatures and voltages. The data obtained can then be used to design and optimize power electronic circuits.

12. 1. 2 Analog Circuits

Transistors used in analog circuits are characterized for their ability to amplify and process signals. The following parameters are of primary interest:

Small-signal current gain (hfe): The ratio of the collector current to the base current when the transistor is operating in the active region.
Input impedance (Zin): The impedance seen at the base of the transistor.
Output impedance (Zout): The impedance seen at the collector of the transistor.
Transconductance (gm): The ratio of the change in collector current to the change in base-emitter voltage.
Cutoff frequency (fT): The frequency at which the small-signal current gain drops to 1.

The characterization of analog transistors typically involves measuring these parameters at different frequencies and signal levels. The data obtained can then be used to design and optimize analog circuits.

12. 1. 3 Digital Circuits

Transistors used in digital circuits are characterized for their ability to switch between on and off states quickly and reliably. The following parameters are of primary interest:

Propagation delay (tpd): The time it takes for a signal to propagate through the transistor.
Noise margin (NM): The difference between the input voltage that causes the transistor to switch and the input voltage that causes it to remain in the same state.
Fan-out: The number of gates that the transistor can drive without significantly affecting its performance.

The characterization of digital transistors typically involves measuring these parameters under various operating conditions, such as different temperatures and voltages. The data obtained can then be used to design and optimize digital circuits.

12. 1. 4 Radio Frequency (RF) Circuits

Transistors used in RF circuits are characterized for their ability to handle high frequencies and maintain good signal integrity. The following parameters are of primary interest:

RF power (PRF): The maximum power that the transistor can handle without damaging itself.
Insertion loss (IL): The reduction in signal power caused by the transistor.
Return loss (RL): The amount of signal power that is reflected back from the transistor.
Noise figure (NF): The amount of noise added to the signal by the transistor.

The characterization of RF transistors typically involves measuring these parameters at different frequencies and power levels. The data obtained can then be used to design and optimize RF circuits.

Conclusion

The characterization of transistors is essential for ensuring their proper operation and reliability in different applications. The specific parameters and techniques used for characterization vary depending on the application. By understanding the different characterization techniques and their relevance to specific applications, engineers can design and optimize electronic devices that meet the desired performance requirements.

12.2 Case Studies of Real-World Failures and their Analysis

The Tacoma Narrows Bridge, spanning the Tacoma Narrows Strait in Washington, United States, gained notoriety for its catastrophic collapse on November 7, 1940, just four months after its opening. The bridge, hailed as an architectural marvel, exhibited a unique susceptibility to aeroelastic flutter, a phenomenon where wind induces self-sustaining oscillations in a structure.

As winds reached speeds of approximately 40 miles per hour, the bridge began to sway violently, with its center span oscillating in an elliptical pattern. The oscillations intensified, causing the bridge to twist and buckle. Within a matter of minutes, the entire center span collapsed into the waters below, taking with it six people who were on the bridge at the time.

An investigation into the collapse revealed a combination of factors that contributed to the failure. The bridge's lightweight, streamlined design, intended to reduce wind resistance, inadvertently made it more susceptible to flutter. Additionally, the bridge's suspension system lacked sufficient damping mechanisms, allowing oscillations to amplify and become uncontrollable.

The Tacoma Narrows Bridge collapse served as a stark reminder of the importance of considering aeroelastic effects in bridge design. Engineers now employ sophisticated computer models and wind tunnel testing to predict and mitigate flutter risks, ensuring the safety and stability of modern bridges.

Case Study 2: The Challenger Space Shuttle Disaster

On January 28, 1986, the Space Shuttle Challenger disintegrated 73 seconds after liftoff, resulting in the tragic loss of all seven crew members. The investigation into the disaster identified a catastrophic failure of the shuttle's solid rocket boosters (SRBs) as the root cause.

During launch, hot gases from the SRBs leaked through a faulty O-ring, a seal designed to prevent leakage. The escaping gases burned through the SRB casing, eventually causing it to rupture. The resulting explosion destroyed the shuttle's external fuel tank and triggered the disintegration of the orbiter.

The Challenger disaster highlighted the critical importance of thorough risk assessment and quality control in aerospace engineering. Engineers now conduct rigorous inspections and testing to ensure the integrity of critical components, minimizing the risk of similar failures in future space missions.

Case Study 3: The Bhopal Gas Tragedy

In the early hours of December 3, 1984, a major industrial disaster occurred at the Union Carbide India Limited (UCIL) pesticide plant in Bhopal, India. A massive release of toxic methyl isocyanate (MIC) gas from the plant's storage tank enveloped the surrounding area, affecting over half a million people.

The tragedy was caused by a combination of factors, including inadequate safety measures, poor maintenance, and negligent operating procedures. The storage tank was overfilled, and a faulty valve allowed MIC to escape into the atmosphere. The plant lacked proper ventilation and warning systems, exacerbating the impact of the gas leak.

The Bhopal Gas Tragedy remains one of the deadliest industrial disasters in history, with estimates of fatalities ranging from 3,787 to 15,000 people. The disaster highlighted the urgent need for stringent regulations and comprehensive safety measures in the chemical industry to prevent similar tragedies from occurring.

Analysis of Real-World Failures

These case studies illustrate the devastating consequences that can arise from engineering failures. However, they also provide valuable lessons that can enhance future design and engineering practices. By thoroughly investigating failures, engineers can identify systemic weaknesses and implement measures to mitigate risks.

Common themes that emerge from real-world failures include:

Inadequate Risk Assessment and Mitigation: Failures often result from underestimating potential risks or

failing to implement appropriate mitigation strategies.
Design and Engineering Flaws: Faulty designs, improper material selection, or inadequate testing can lead to catastrophic failures.
Lack of Quality Control and Maintenance: Negligence in manufacturing, inspection, or maintenance can compromise the integrity of structures or systems.
Human Error and Organizational Factors: Errors made by individuals or systemic organizational issues can contribute to failures.
Unforeseen Conditions: Failures can occur when designs fail to account for extreme or unexpected conditions.

By understanding these themes and implementing proactive measures, engineers can strive to prevent similar failures in the future, ensuring the safety and reliability of our infrastructure, products, and systems.

Chapter 13: Standards and Guidelines

13.1 Industry Standards for Transistor Testing

Industry standards provide a common framework for transistor testing, ensuring consistency and accuracy in the evaluation of these critical components.

IEEE. JEDEC Standards

The Institute of Electrical and Electronics Engineers (IEEE) and the Joint Electron Device Engineering Council (JEDEC) have jointly developed a comprehensive set of standards for transistor testing. These standards cover a wide range of aspects, including:

Measurement methods: Define standardized procedures for measuring key transistor parameters, such as current gain, transconductance, and threshold voltage.
Test conditions: Specify the environmental conditions, such as temperature and bias voltage, under which transistors are to be tested.
Data reporting: Establish guidelines for reporting test results, ensuring that data is presented in a consistent and meaningful manner.

Benefits of IEEE. JEDEC Standards

Adherence to IEEE. JEDEC standards offers several benefits for the electronics industry:

Improved comparability: Standards enable the comparison of transistor performance across different manufacturers and devices.
Enhanced reliability: Standardized testing procedures help ensure that transistors meet the specified performance criteria, reducing the risk of device failures.
Increased productivity: Standardized methods streamline the testing process, reducing the time and resources required for evaluation.
Global acceptance: IEEE. JEDEC standards are recognized worldwide, facilitating international trade and cooperation.

Other Industry Standards

In addition to IEEE. JEDEC standards, other organizations have also developed industry standards for transistor testing. These include:

ASTM International (ASTM): Develops standards for a variety of materials, including semiconductors. ASTM standards for transistor testing cover aspects such as packaging and handling.
International Electrotechnical Commission (IEC): Publishes international standards for electrical and

electronic devices. IEC standards for transistor testing address issues such as safety and performance testing.

Importance of Compliance

Compliance with industry standards is essential for several reasons:

Quality assurance: Standards help ensure that transistors meet the required specifications and perform reliably.
Customer satisfaction: Adherence to standards demonstrates a commitment to quality and customer satisfaction.
Regulatory compliance: Some industries, such as automotive and medical, have specific regulations that require compliance with certain standards.

Conclusion

Industry standards for transistor testing provide a common framework for evaluating these critical electronic components. Adherence to these standards promotes comparability, reliability, productivity, and global acceptance. By following established testing procedures and reporting guidelines, manufacturers and users can ensure the consistent and accurate assessment of transistor performance, ultimately contributing to the development of high-quality electronic devices.

13.2 Best Practices and Guidelines for Characterization

Characterization plays a crucial role in literature, creating memorable and relatable characters that drive the plot and engage readers. To ensure effective characterization, authors should adhere to established best practices and guidelines.

1. Establish Clear Goals and Motivations:

Begin by defining the character's purpose within the story. Determine their goals, motivations, and desires. This foundation will guide the development of their actions, dialogue, and interactions with others.

2. Craft a Compelling Backstory:

Every character has a past that influences their present. Create a rich backstory that provides context for their motivations, beliefs, and personality traits. Consider their childhood experiences, relationships, and significant events that have shaped their character.

3. Develop Physical and Emotional Characteristics:

Describe the character's physical appearance, including their height, weight, hair color, eye color, and distinctive features. This helps readers visualize the character and connect with their physical presence. Additionally, explore their emotional depth by identifying their strengths, weaknesses, fears, and vulnerabilities.

4. Use Dialogue and Action to Reveal Character:

Dialogue provides a powerful tool for revealing a character's thoughts, feelings, and personality. Create authentic conversations that showcase their speech patterns, vocabulary, and mannerisms. Similarly, describe actions and reactions that demonstrate their values, beliefs, and responses to different situations.

5. Create Conflict and Challenges:

Compelling characters face challenges and conflicts that test their limits. Introduce obstacles, obstacles, and dilemmas that force them to make difficult decisions, grow, and evolve. These challenges should align with the character's goals and motivations, creating a sense of tension and suspense.

6. Show Growth and Development:

Characters should not remain static throughout the story. Allow them to experience growth and development as they navigate challenges, learn from their mistakes, and adapt to changing circumstances. This growth can be gradual or sudden, but it should be believable and driven by the events of the story.

7. Ensure Consistency and Avoid Stereotypes:

Maintain consistency in the character's behavior, actions,

and speech patterns throughout the story. Avoid falling into stereotypes or creating caricatures. Strive for originality and depth, making each character unique and memorable.

8. Seek Feedback and Revise:

Seek feedback from beta readers, critique partners, or writing workshops to gain insights into the effectiveness of your characterization. Revise and refine the characters based on feedback, ensuring they resonate with readers and support the overall narrative.

Additional Guidelines:

Consider Perspective: Determine whether the character will be revealed through first-person or third-person narration. Each perspective offers different insights and challenges.
Balance Complexity and Accessibility: Create characters that are complex and multidimensional, but avoid overwhelming readers with excessive detail or psychological depth.
Respect Diversity: Represent a diverse range of characters with different backgrounds, identities, and experiences. Avoid stereotypes or offensive portrayals.
Use Symbolism and Metaphor: Utilize symbolism, metaphors, and imagery to enrich the characterization. Connect physical or emotional characteristics to larger themes or ideas.
Focus on the Reader's Experience: Ultimately, the goal of

characterization is to engage and entertain readers. Craft characters that resonate with readers on an emotional level and leave a lasting impression.

Chapter 14: Safety and Ethical Considerations

14.1 Safety Precautions for Handling and Testing Transistors

Transistors are semiconductor devices that play a vital role in electronic circuits. They are used in a wide range of applications, from small-scale devices such as calculators and watches to large-scale systems such as computers and telecommunication networks. While transistors are generally safe to handle and test, there are certain precautions that should be taken to ensure the safety of both the user and the device.

Electrostatic Discharge (ESD)

One of the most important safety precautions when handling and testing transistors is to protect them from electrostatic discharge (ESD). ESD occurs when two objects with different electrical charges come into contact, resulting in a sudden flow of electrons from the charged object to the grounded object. This can damage or even destroy sensitive electronic components such as transistors.

To prevent ESD, it is important to ground yourself and any equipment you are using before handling transistors.

You can do this by touching a metal surface or by wearing an ESD wrist strap. You should also avoid touching the terminals of transistors with your bare hands.

Heat

Transistors can also be damaged by heat. When a transistor is operating, it generates heat as a byproduct of its normal operation. If the transistor is not properly cooled, the heat can build up and cause the transistor to fail.

To prevent overheating, transistors should be mounted on heat sinks. Heat sinks are metal devices that help to dissipate heat away from transistors. You should also make sure that the transistors are properly ventilated.

Overvoltage

Overvoltage is another potential hazard when handling and testing transistors. Overvoltage occurs when the voltage applied to a transistor exceeds the maximum voltage that the transistor can withstand. This can damage or even destroy the transistor.

To prevent overvoltage, it is important to use a power supply that is capable of providing a stable voltage. You should also make sure that the voltage applied to the transistor does not exceed the maximum voltage specified in the transistor's datasheet.

Short Circuits

Short circuits occur when two terminals of a transistor are accidentally connected together. This can cause the transistor to draw excessive current, which can damage or even destroy the transistor.

To prevent short circuits, it is important to be careful when handling and testing transistors. You should avoid touching the terminals of transistors with your bare hands or with metal objects. You should also make sure that the transistors are properly insulated.

Chemical Hazards

Some transistors contain hazardous chemicals, such as lead and mercury. It is important to handle and dispose of these transistors properly to avoid exposure to these chemicals.

You should always wear gloves when handling transistors that contain hazardous chemicals. You should also avoid breathing in the fumes from these transistors. If you do come into contact with hazardous chemicals, wash your hands thoroughly with soap and water.

By following these safety precautions, you can help to ensure the safety of both yourself and the transistors you are handling and testing.

14.2 Ethical Considerations in Data Collection and Interpretation

Data collection and interpretation are essential components of research, but they also present ethical challenges that researchers must navigate responsibly. Ethical considerations arise from the potential for data collection to intrude on privacy, cause harm to participants, or lead to biased or discriminatory outcomes.

Privacy and Confidentiality

One of the primary ethical concerns in data collection is protecting the privacy and confidentiality of participants. Personal information collected through surveys, interviews, or observations must be handled with care to prevent unauthorized access or disclosure. Researchers have a responsibility to:

Obtain informed consent from participants, ensuring they understand the purpose of the research, how their data will be used, and how their privacy will be protected.
Securely store and transmit data to prevent breaches or unauthorized use.
Limit data sharing to authorized researchers or organizations, with appropriate safeguards in place.
Anonymize or de-identify data whenever possible, removing personally identifiable information.

Participant Harm

Data collection methods should not cause physical, psychological, or emotional harm to participants. Researchers must:

Minimize potential risks and discomfort during data collection procedures.
Ensure participants are fully informed of the potential risks and benefits of participating.
Provide appropriate support and referral services to participants who experience distress or discomfort.
Respect cultural and ethical sensitivities when conducting research, avoiding practices that could be perceived as disrespectful or offensive.

Bias and Discrimination

Data collection methods and interpretation can introduce bias and discrimination, leading to inaccurate or unfair conclusions. Researchers must:

Use representative sampling techniques to avoid biases in the selection of participants.
Design data collection instruments to minimize potential biases in the questions or tasks presented.
Analyze data carefully to identify and address any biases that may have influenced the results.
Avoid making generalizations beyond the specific population studied, acknowledging the limitations of the data.
Consider the potential impact of data on marginalized or vulnerable populations, ensuring that findings are

interpreted in a way that promotes equity and inclusion.

Transparency and Accountability

Researchers have a responsibility to be transparent about their data collection and interpretation methods. This includes:

Disclosing potential conflicts of interest that may bias the research.
Clearly documenting the research design, methodology, and data analysis techniques used.
Allowing independent researchers to review and verify the research findings.
Acknowledging the limitations and uncertainties in the data and its interpretation.

Ethical Review Boards

Many institutions have ethical review boards (ERBs) or institutional review boards (IRBs) that review research proposals to ensure ethical compliance. ERBs typically assess the potential risks and benefits of the research, the informed consent process, and the protection of participant privacy and confidentiality. Researchers are required to submit their research plans for IRB review and approval before data collection begins.

Consequences of Ethical Violations

Ethical violations in data collection and interpretation can have serious consequences for researchers, participants, and the scientific community. These consequences may include:

Loss of funding and reputation.
Retraction of published research.
Legal action from participants or regulatory bodies.
Damage to the trust between researchers and the public.

Conclusion

Ethical considerations are paramount in data collection and interpretation. Researchers must prioritize privacy, minimize harm, avoid bias and discrimination, ensure transparency, and adhere to ethical review processes. By adhering to these principles, researchers can conduct ethical research that contributes to scientific knowledge and the well-being of society.

Chapter 15: Career Paths and Opportunities

15.1 Careers in Transistor Testing and Characterization

The semiconductor industry is a rapidly growing field that offers a variety of exciting career opportunities. One area of particular interest is transistor testing and characterization. Transistors are essential components in electronic devices, and their performance must be carefully tested and characterized to ensure that they meet the required specifications.

What is transistor testing and characterization.

Transistor testing and characterization involves measuring the electrical properties of transistors to ensure that they meet the desired specifications. This process can be performed manually or using automated equipment. Manual testing is typically used for small batches of transistors, while automated testing is used for large-volume production.

What are the different types of transistor tests.

There are a variety of different transistor tests that can be performed, depending on the specific requirements of the application. Some of the most common tests include:

DC tests: These tests measure the static electrical properties of transistors, such as their forward and reverse bias characteristics.
AC tests: These tests measure the dynamic electrical properties of transistors, such as their frequency response and gain.
Environmental tests: These tests measure the performance of transistors under different environmental conditions, such as temperature, humidity, and vibration.

What are the different career opportunities in transistor testing and characterization.

There are a variety of different career opportunities available in transistor testing and characterization. Some of the most common positions include:

Test engineer: Test engineers are responsible for developing and executing test plans for transistors. They also analyze test results and identify any potential problems.
Characterisation engineer: Characterisation engineers are responsible for measuring the electrical properties of transistors and developing models that can be used to predict their performance in different applications.
Failure analysis engineer: Failure analysis engineers are responsible for investigating the causes of transistor

failures. They use a variety of techniques to identify the root cause of the failure and recommend corrective actions.

What are the educational requirements for a career in transistor testing and characterization.

A bachelor's degree in electrical engineering or a related field is typically required for a career in transistor testing and characterization. Some employers may also require a master's degree or a PhD.

What are the salary expectations for a career in transistor testing and characterization.

The salary expectations for a career in transistor testing and characterization vary depending on the specific position, experience, and location. According to the U. S. Bureau of Labor Statistics, the median annual salary for electrical engineers was $100,420 in May 2021.

What are the job outlook for a career in transistor testing and characterization.

The job outlook for a career in transistor testing and characterization is expected to be good over the next few years. The increasing demand for electronic devices is expected to drive the need for more transistors, which will in turn lead to more jobs in transistor testing and characterization.

How can I get started in a career in transistor testing and characterization.

There are a number of ways to get started in a career in transistor testing and characterization. Some of the most common ways include:

Get a bachelor's degree in electrical engineering or a related field.
Gain experience in transistor testing and characterization through internships or co-ops.
Network with people in the semiconductor industry.
Attend industry conferences and workshops.

What are the benefits of a career in transistor testing and characterization.

There are a number of benefits to a career in transistor testing and characterization, including:

Good salary potential.
Job security.
Opportunities for advancement.
Challenging and rewarding work.

If you are interested in a career in transistor testing and characterization, there are a number of resources available to help you get started. The Semiconductor Industry Association (SIA) is a good place to start your search for information about the industry and job opportunities.

15.2 Educational Requirements and Professional Development

The educational requirements for special education teachers vary from state to state, but most require a bachelor's degree in special education or a related field. Some states also require teachers to have a master's degree in special education or a related field. In addition to their coursework, special education teachers must also complete a state-approved teacher preparation program. This program typically includes coursework in special education methods, curriculum development, assessment, and behavior management.

Professional Development for Special Education Teachers

Professional development is an essential part of being a special education teacher. Special education teachers must stay up-to-date on the latest research and best practices in order to provide the best possible education for their students. There are many different ways for special education teachers to participate in professional development, including attending workshops, conferences, and online courses.

Why Professional Development is Important for Special Education Teachers

There are many reasons why professional development is important for special education teachers. First, professional development helps teachers to stay up-to-date on the latest research and best practices in special education. This knowledge can help teachers to improve their teaching methods and provide better instruction for their students. Second, professional development helps teachers to develop new skills and strategies that they can use in their classrooms. This can help teachers to become more effective educators and improve the learning outcomes of their students. Third, professional development helps teachers to network with other professionals in the field of special education. This can help teachers to share ideas and resources, and it can also help them to stay informed about the latest trends in special education.

Conclusion

Educational requirements and professional development are essential for special education teachers. Special education teachers must have the knowledge and skills to provide the best possible education for their students. By completing a rigorous educational program and participating in ongoing professional development, special education teachers can ensure that they are providing their students with the best possible education.

Here are some additional tips for special education teachers who are looking to enhance their professional development:

Develop a professional development plan. This plan should

outline your goals for professional development and the steps you will take to achieve those goals.
Attend workshops, conferences, and online courses. This is a great way to learn about new research and best practices in special education.
Network with other special education professionals. This can help you to share ideas and resources, and it can also help you to stay informed about the latest trends in special education.
Read professional journals and books. This is a great way to stay up-to-date on the latest research and best practices in special education.
Participate in research projects. This can help you to learn about new research methods and findings, and it can also help you to contribute to the field of special education.

By following these tips, special education teachers can enhance their professional development and provide the best possible education for their students.

Chapter 16: Research and Development

16.1 Current Research Trends in Transistor Testing

As the demand for faster, more efficient, and smaller electronic devices continues to grow, the need for advanced transistor testing techniques becomes paramount. This one delves into the cutting-edge research trends that are transforming the field of transistor testing, paving the way for future advancements in electronics.

1. Non-Destructive Testing Techniques

Traditional transistor testing methods often involve destructive procedures that damage the device under test. Non-destructive testing (NDT) techniques, on the other hand, enable the evaluation of transistor performance without compromising its integrity. These methods are gaining increasing traction due to their ability to provide valuable insights into the device's operation under real-world conditions.

One promising NDT technique is non-contact electrical impedance spectroscopy (EIS). This approach analyzes the electrical impedance of the transistor to extract information about its physical properties, such as carrier concentration and mobility. By monitoring the impedance changes over time or under different operating conditions, researchers can identify subtle defects or performance degradation that would otherwise remain undetected.

2. High-Throughput Testing

The proliferation of electronic devices in various applications has created a pressing need for high-throughput transistor testing. Conventional testing methods, which involve time-consuming manual or semi-automated processes, are becoming increasingly impractical for evaluating large volumes of transistors.

Automated testing systems, equipped with advanced instrumentation and parallel testing capabilities, are addressing this challenge. These systems leverage sophisticated algorithms to optimize test procedures, reduce test times, and improve overall efficiency. The use of artificial intelligence (AI) and machine learning (ML) techniques is further enhancing the accuracy and reliability of high-throughput transistor testing.

3. Wide-Bandgap Semiconductor Transistors

Wide-bandgap (WBG) semiconductors, such as gallium nitride (GaN) and silicon carbide (SiC), are revolutionizing transistor design. These materials offer superior properties, including higher breakdown voltage, lower on-state resistance, and faster switching speeds compared to conventional silicon transistors.

Testing WBG transistors requires specialized techniques due to their unique electrical characteristics. Researchers are exploring innovative approaches, such as using pulsed measurement systems and developing high-frequency test fixtures, to accurately characterize the performance of these advanced devices.

4. Power Transistor Testing

Power transistors are essential components in high-power electronic systems, such as electric vehicles, industrial drives, and renewable energy applications. Testing power transistors involves unique challenges due to their high current and voltage handling capabilities.

Researchers are focusing on developing novel testing methodologies that can evaluate the reliability and performance of power transistors under harsh operating conditions. These methods include accelerated stress testing, thermal impedance characterization, and failure analysis techniques.

5. Compact Modeling and Test Data Analytics

Compact modeling plays a vital role in designing and simulating transistor circuits. Accurate and efficient models are essential for predicting device behavior under various operating conditions and optimizing circuit performance.

Recent research trends in compact modeling focus on developing physics-based models that incorporate the latest understanding of transistor physics. These models enable more accurate simulations and reduce the need for extensive experimental characterization.

Additionally, test data analytics is gaining momentum as a valuable tool for extracting insights from transistor testing results. Machine learning algorithms can analyze large volumes of test data to identify patterns, detect anomalies, and predict transistor performance over time. This information can be leveraged to optimize testing procedures, improve yield, and enhance device reliability.

Conclusion

The field of transistor testing is undergoing a transformative phase, driven by the relentless pursuit of advanced electronic technologies. Non-destructive testing, high-throughput testing, wide-bandgap semiconductor transistors, power transistor testing, and compact modeling and test data analytics represent the key research trends that are shaping the future of transistor testing. By embracing these cutting-edge techniques, researchers and engineers are paving the way for the development of more efficient, reliable, and high-performance electronic devices.

16.2 Future Research Opportunities and Challenges

The field of machine learning offers a vast array of exciting research opportunities. One promising area is the development of new algorithms that can handle increasingly complex data types, such as images, videos, and natural language. Another area of active research is the development of more efficient and scalable algorithms, which can be applied to larger datasets and more complex problems.

In addition, there is a growing interest in the development of interpretable machine learning models. These models can provide insights into the decision-making process of machine learning algorithms, making them more trustworthy and reliable.

Another important research area is the development of new methods for evaluating the performance of machine learning models. Traditional evaluation metrics, such as accuracy and precision, may not be adequate for assessing the performance of models on complex tasks, such as natural language understanding and computer vision.

Challenges in Machine Learning

While machine learning offers great promise, there are also a number of challenges that need to be addressed. One challenge is the lack of interpretability of machine learning models. This can make it difficult to understand why models make certain decisions, which can limit their trustworthiness and reliability.

Another challenge is the potential for bias in machine learning models. This can occur when the training data is not representative of the population that the model will be used on. Bias can lead to unfair or inaccurate predictions, which can have serious consequences.

To conclude, there is the challenge of scalability. Machine learning models can be computationally expensive to train and deploy, especially for large datasets. This can limit their applicability to real-world problems.

Overcoming the Challenges

The challenges facing machine learning are significant, but they are not insurmountable. Researchers are actively working on new methods to address these challenges, and there has been significant progress in recent years.

For example, researchers have developed new techniques for making machine learning models more interpretable. These techniques can provide insights into the decision-making process of models, making them more trustworthy and reliable.

Researchers are also working on new methods to reduce bias in machine learning models. These methods can help to ensure that models are fair and accurate, even when the training data is not representative of the population that the model will be used on.

Finally, researchers are working on new methods to make machine learning models more scalable. These methods can help to reduce the computational cost of training and deploying models, making them more applicable to real-world problems.

The Future of Machine Learning

Machine learning is a rapidly evolving field, with new developments emerging all the time. The challenges facing machine learning are significant, but researchers are making progress in addressing them. As these challenges are overcome, machine learning will become increasingly powerful and versatile, and it will have a transformative impact on a wide range of industries and applications.

Chapter 17: Conclusion

17.1 Importance of Continuous Improvement in Transistor Testing

Transistors are essential components in electronic devices, and their performance and reliability directly impact the overall functionality of the device. To ensure that transistors meet the required specifications and perform optimally, rigorous testing is crucial. Continuous improvement in transistor testing is of paramount importance for several reasons:

1. Evolving Device Technologies:

The semiconductor industry is constantly evolving, with new transistor designs and technologies emerging at a rapid pace. Continuous improvement in testing methods and techniques is necessary to keep pace with these advancements and ensure that transistors are tested accurately and effectively. Failure to adapt to these changes can result in inaccurate or incomplete testing, leading to potential device failures.

2. Increasing Complexity and Functionality:

Modern transistors exhibit increasing complexity and

functionality, incorporating multiple features and operating modes. This complexity demands more sophisticated testing approaches to evaluate all aspects of transistor performance and ensure their proper operation under various conditions. Continuous improvement in testing methods enables the development of tests that can capture the nuances of these complex devices.

3. Enhanced Reliability and Yield:

Thorough and reliable testing is essential for identifying and eliminating defective transistors, improving overall device yield and reliability. Continuous improvement in testing techniques helps minimize the chances of undetected defects, reducing the risk of device failures and enhancing the overall quality of electronic products.

4. Cost Reduction and Optimization:

Efficient and accurate testing can significantly reduce manufacturing costs. By identifying defective transistors early in the production process, continuous improvement in testing helps minimize waste and rework, optimizing the manufacturing process and reducing production costs.

5. Safety and Compliance:

In certain applications, such as medical devices and automotive electronics, the proper functioning of transistors is critical for ensuring safety and compliance with regulatory standards. Continuous improvement in testing ensures that transistors meet the stringent requirements of these applications, minimizing the risks associated with device failures.

6. Industry Best Practices:

Continuous improvement in transistor testing aligns with industry best practices and standards. By adopting the latest testing methods and techniques, manufacturers demonstrate their commitment to quality and reliability, enhancing their reputation and competitiveness in the market.

7. Competitive Advantage:

Companies that invest in continuous improvement of transistor testing gain a competitive advantage by delivering high-quality products with improved performance and reliability. This can lead to increased customer satisfaction, repeat business, and brand recognition. By embracing this approach, manufacturers can ensure the accuracy and effectiveness of transistor testing, keeping pace with evolving device technologies, enhancing device reliability, reducing costs, and maintaining industry best practices. Ultimately, continuous improvement leads to the production of high-quality electronic devices that meet the demands of modern applications and consumer expectations.

17.2 Future Outlook and Emerging Technologies

The realm of technology is perpetually evolving, and the future holds an abundance of promising prospects. As we delve into the upcoming years, several emerging technologies are poised to revolutionize various aspects of our lives, reshaping industries and opening up unprecedented possibilities.

Artificial Intelligence (AI)

Artificial intelligence (AI) has emerged as a transformative force across numerous sectors. Its capabilities extend beyond automating repetitive tasks; AI algorithms can now analyze vast datasets, identify patterns, and make predictions with remarkable accuracy. This has far-reaching implications for fields such as healthcare, finance, and manufacturing, where AI-driven systems can enhance decision-making, improve efficiency, and unlock new avenues for innovation.

Blockchain Technology

Blockchain technology, the foundation of cryptocurrencies like Bitcoin, has gained significant traction. Its decentralized and immutable nature offers unique advantages for secure data storage and transactions. Blockchain-based applications are proliferating in industries such as supply chain management, voting systems, and healthcare, providing enhanced transparency, traceability, and security.

Quantum Computing

Quantum computing harnesses the principles of quantum mechanics to perform computations exponentially faster than conventional computers. This has the potential to accelerate scientific research, drug discovery, and materials science. Quantum algorithms can tackle complex problems that are currently intractable, unlocking new frontiers of knowledge and innovation.

Augmented and Virtual Reality (AR. VR)

Augmented reality (AR) and virtual reality (VR) technologies are blurring the lines between the physical and digital worlds. AR superimposes digital information onto the real world, enhancing user experiences in fields such as education, healthcare, and retail. VR creates immersive virtual environments, enabling novel forms of entertainment, training, and collaboration.

Internet of Things (IoT)

The Internet of Things (IoT) connects everyday objects to the internet, enabling them to communicate and exchange data. This has spawned a plethora of applications in smart homes, smart cities, and industrial automation. IoT devices can monitor environmental conditions, control appliances, and provide valuable insights into usage patterns, leading to improved efficiency and convenience.

Robotics and Automation

Robotics and automation are transforming the manufacturing, healthcare, and service industries. Robots are becoming increasingly sophisticated, capable of performing complex tasks with precision and efficiency. Automation technologies are streamlining processes, reducing costs, and enhancing productivity.

5G Connectivity

The advent of 5G connectivity promises significantly faster speeds, lower latency, and increased capacity. This will fuel the development of new applications, such as self-driving cars, smart cities, and immersive entertainment experiences. 5G networks will serve as the backbone for the future digital ecosystem, enabling ubiquitous connectivity and real-time data processing.

Biotechnology and Genetic Engineering

Biotechnology and genetic engineering hold immense potential to revolutionize healthcare and agriculture. Advancements in gene editing techniques, such as CRISPR-Cas9, are enabling scientists to manipulate the genetic code of organisms with unprecedented precision. This has implications for treating genetic diseases, developing new therapies, and improving crop yields.

Nanotechnology

Nanotechnology deals with the manipulation of matter at the atomic and molecular scale. Nanomaterials possess unique properties that can be harnessed for various applications, including advanced materials, drug delivery systems, and energy storage. Nanotechnology has the potential to drive innovation in fields as diverse as electronics, medicine, and manufacturing.

Conclusion

The future of technology is brimming with possibilities. Emerging technologies like AI, blockchain, quantum computing, AR. VR, IoT, robotics, 5G connectivity, biotechnology, and nanotechnology are poised to reshape our world in profound ways. By embracing these advancements, we can unlock new frontiers of knowledge, drive economic growth, and improve the quality of life for generations to come. However, it is crucial to approach these technologies with a responsible and ethical mindset, ensuring that they are deployed in ways that benefit humanity as a whole.

Made in the USA
Coppell, TX
24 November 2024